HALSALL    10

# Fundamentals of HVAC Systems

# Fundamentals of HVAC Systems

Prepared by

**Robert McDowall,** P. Eng.
Engineering Change Inc.

American Society of Heating, Refrigerating and
Air-Conditioning Engineers Inc.
1791 Tullie Circle NE, Atlanta, GA 30329, USA

ELSEVIER

AMSTERDAM • BOSTON • HEIDELBERG • LONDON
NEW YORK • OXFORD PARIS • SAN DIEGO
SAN FRANCISCO • SINGAPORE • SYDNEY • TOKYO
Butterworth-Heinemann is an imprint of Elsevier

Butterworth-Heinemann is an imprint of Elsevier
Linacre House, Jordan Hill, Oxford OX2 8DP, UK
The Boulevard, Langford Lane, Kidlington, Oxford OX5 1GB, UK
84 Theobald's Road, London WC1X 8RR, UK
Radarweg 29, PO Box 211, 1000 AE Amsterdam, The Netherlands
30 Corporate Drive, Suite 400, Burlington, MA 01803, USA
525 B Street, Suite 1900, San Diego, CA 92101-4495, USA

First edition 2006

Copyright © 2006, American Society of Heating, Refrigerating and Air-Conditioning
Engineers, Inc. and Elsevier Inc Published by Elsevier 2006. All rights reserved

The right of American Society of Heating, Refrigerating and Air-Conditioning
Engineers, Inc. and Elsevier Inc to be identified as the author of this work has been
asserted in accordance with the Copyright, Designs and Patents Act 1988

No part of this publication may be reproduced, stored in a retrieval system
or transmitted in any form or by any means electronic, mechanical, photocopying,
recording or otherwise without the prior written permission of the publisher

Permissions may be sought directly from Elsevier's Science and  e hnology Rights
Department in Oxfor   UK. Phone: (44) 1865 843830,   ax: (44) 1865 853333, e-mail:
permissions@elsevier.co.uk.  ou may also complete your request on-line via the
Elsevier homepage: http://www.elsevier.com b  selecting "Customer Support" and
then "Obtaining  ermissions".

Notice
No responsibility is assumed by the publisher for any injury and/or damage to persons
or property as a matter of products liability, negligence or otherwise, or from any use
or operation of any methods, products, instructions or ideas contained in the material
herein. Because of rapid advances in the medical sciences, in particular, independent
verification of diagnoses and drug dosages should be made

**British Library Cataloguing in Publication Data**
A catalogue record for this book is available from the British Library

**Library of Congress Cataloging-in-Publication Data**
A catalog record for this book is available from the Library of Congress

ISBN–13: 978-0-12-372497-7
ISBN–10: 0-12-372497-X

For information on all Butterworth-Heinemann publications
visit our web site at books.elsevier.com

Printed and bound in Great Briatin
06  07  08  09  10   10  9  8  7  6  5  4  3  2

# Working together to grow
## libraries in developing countries

www.elsevier.com | www.bookaid.org | www.sabre.org

ELSEVIER    BOOK AID International    Sabre Foundation

# Contents

| | |
|---|---|
| Foreword | ix |
| **1 Introduction to HVAC** | **1** |
| Study Objectives of Chapter 1 | 1 |
| 1.1 Introduction | 1 |
| 1.2 Brief History of HVAC | 2 |
| 1.3 Scope of Modern HVAC | 3 |
| 1.4 Introduction to Air-conditioning Processes | 3 |
| 1.5 Objective: What is your system to achieve? | 4 |
| 1.6 Environment For Human Comfort | 6 |
| The Next Step | 8 |
| Summary | 8 |
| Bibliography | 9 |
| **2 Introduction to HVAC Systems** | **10** |
| Study Objectives of Chapter 2 | 10 |
| 2.1 Introduction | 10 |
| 2.2 Introducing the Psychrometric Chart | 11 |
| 2.3 Basic Air-Conditioning System | 20 |
| 2.4 Zoned Air-Conditioning Systems | 23 |
| 2.5 Choosing an Air-Conditioning System | 26 |
| 2.6 System Choice Matrix | 28 |
| The Next Step | 30 |
| Summary | 30 |
| Bibliography | 31 |
| **3 Thermal Comfort** | **32** |
| Study Objectives of Chapter 3 | 32 |
| 3.1 Introduction: What is Thermal Comfort? | 32 |
| 3.2 Seven Factors Influencing Thermal Comfort | 33 |
| 3.3 Conditions for Comfort | 36 |
| 3.4 Managing Under Less Than Ideal Conditions | 39 |
| 3.5 Requirements of Non-Standard Groups | 40 |
| The Next Step | 41 |
| Summary | 41 |
| Bibliography | 42 |

vi   Contents

### 4  Ventilation and Indoor Air Quality    43

Study Objectives of Chapter 4    43
4.1  Introduction    43
4.2  Air Pollutants and Contaminants    44
4.3  Indoor Air Quality Effects on Health and Comfort    45
4.4  Controlling Indoor Air Quality    47
4.5  ASHRAE Standard 62 Ventilation for Acceptable Indoor Air Quality    52
The Next Step    58
Summary    58
Bibliography    59

### 5  Zones    60

Study Objectives of Chapter 5    60
5.1  Introduction    60
5.2  What is a Zone?    61
5.3  Zoning Design    62
5.4  Controlling the Zone    65
The Next Step    67
Summary    67

### 6  Single Zone Air Handlers and Unitary Equipment    68

Study Objectives of Chapter 6    68
6.1  Introduction    68
6.2  Examples of Buildings with Single-zone Package Air-Conditioning Units    69
6.3  Air-Handling Unit Components    70
6.4  Refrigeration Equipment    75
6.5  System Performance Requirements    80
6.6  Rooftop Units    82
6.7  Split Systems    85
The Next Step    86
Summary    86
Bibliography    87

### 7  Multiple Zone Air Systems    88

Study Objectives of Chapter 7    88
7.1  Introduction    88
7.2  Single-Duct, Zoned Reheat, Constant Volume Systems    90
7.3  Single-Duct, Variable Air Volume Systems    92
7.4  By-Pass Box Systems    94
7.5  Constant Volume Dual-Duct, All-Air Systems    95
7.6  Multizone Systems    98
7.7  Three-deck Multizone Systems    99
7.8  Dual-Duct, Variable Air Volume Systems    99
7.9  Dual Path Outside Air Systems    100
The Next Step    101
Summary    101

## 8 Hydronic Systems — 103

Study Objectives of Chapter 8 — 103
8.1 Introduction — 103
8.2 Natural Convection and Low Temperature Radiation Heating Systems — 104
8.3 Panel Heating and Cooling — 108
8.4 Fan Coils — 109
8.5 Two Pipe Induction Systems — 112
8.6 Water Source Heat Pumps — 113
The Next Step — 115
Summary — 115
Bibliography — 116

## 9 Hydronic System Architecture — 117

Study Objectives of Chapter 9 — 117
9.1 Introduction — 117
9.2 Steam — 118
9.3 Water Systems — 120
9.4 Hot Water — 124
9.5 Chilled Water — 127
9.6 Condenser Water — 129
The Next Step — 131
Summary — 131
Bibliography — 132

## 10 Central Plants — 133

Study Objectives of Chapter 10 — 133
10.1 Introduction — 133
10.2 Central Plant Versus Local Plant in a Building — 134
10.3 Boilers — 136
10.4 Chillers — 139
10.5 Cooling Towers — 142
The Next Step — 145
Summary — 145
Bibliography — 147

## 11 Controls — 148

Study Objectives of Chapter 11 — 148
11.1 Introduction — 148
11.2 Basic Control — 150
11.3 Typical Control loops — 155
11.4 Introduction to Direct Digital Control, DDC — 157
11.5 Direct Digital Control of an Air-Handler — 161
11.6 Architecture and Advantages of Direct Digital Controls — 165
The Next Step — 169
Summary — 169
Bibliography — 170

## 12  Energy Conservation Measures — 171

Study Objectives of Chapter 12 — 171
12.1 Introduction — 172
12.2 Energy Considerations for Buildings — 172
12.3 ASHRAE/IESNA Standard 90.1 — 176
12.4 Heat Recovery — 179
12.5 Air-Side and Water-Side Economizers — 183
12.6 Evaporative Cooling — 185
12.7 Control of Building Pressure — 186
The Final Step — 187
Summary — 187
Bibliography — 189

## 13  Special Applications — 190

Study Objectives of Chapter 13 — 190
13.1 Introduction — 190
13.2 Radiant Heating and Cooling Systems — 191
13.3 Thermal Storage Systems — 194
13.4 The Ground as Heat Source and Sink — 204
13.5 Occupant Controlled Windows with HVAC — 206
13.6 Room Air Distribution Systems — 207
13.7 Decoupled or Dual Path, and Dedicated Outdoor Air Systems — 211
Summary — 213
Your Next Step — 215
Bibliography — 216
Index — 219

# Foreword

Every author knows that books are not created in a vacuum, so it is important to acknowledge the support of those who also contributed to the success of the project.

First I'd like to thank my wife Jo-Anne McDowall, who helped with the development of the project, and who read every word, to make sure that a neophyte to the field of HVAC would understand the concepts as they were introduced. Jo-Anne also wrote the chapter summaries and leant her proof-reading and text-editing eye to this task.

I'd like to thank the members of ASHRAE Winnipeg, and especially Bert Phillips, P. Eng., who encouraged and supported me in the development of the project.

Of course, the project would never have come to fruition without ASHRAE members who acted as reviewers.

Finally thanks to ASHRAE and Elsevier staff who made it happen.

Robert McDowall

# Chapter 1
# Introduction to HVAC

## Contents of Chapter 1

Study Objectives of Chapter 1
1.1 Introduction
1.2 History of HVAC
1.3 Scope of Modern HVAC
1.4 Introduction To Air-Conditioning Processes
1.5 Objective: what is your system to achieve?
1.6 Environment For Human Comfort
The Next Step
Summary
Bibliography

## Study Objectives of Chapter 1

Chapter 1 introduces the history, uses and main processes of heating, ventilating and air conditioning. There are no calculations to be done. The ideas will be addressed in detail in later chapters. After studying the chapter, you should be able to:

Define heating, ventilating and air conditioning.
Describe the purposes of heating, ventilating and air conditioning.
Name and describe seven major air-conditioning processes.
Identify five main aspects of a space that influence an occupant's comfort.

## 1.1 Introduction

Heating, Ventilating and Air Conditioning, HVAC, is a huge field. HVAC systems include a range from the simplest hand-stoked stove, used for comfort heating, to the extremely reliable total air-conditioning systems found in submarines and space shuttles. Cooling equipment varies from the small domestic unit to refrigeration machines that are 10,000 times the size, which are used in industrial processes.

Depending on the complexity of the requirements, the HVAC designer must consider many more issues than simply keeping temperatures comfortable. This chapter will introduce you to the fundamental concepts that are used by designers to make decisions about system design, operation, and maintenance.

## 1.2 Brief History of HVAC

For millennia, people have used fire for heating. Initially, the air required to keep the fire going ensured adequate ventilation for the occupants. However, as central furnaces with piped steam or hot water became available for heating, the need for separate ventilation became apparent. By the late 1880s, rules of thumb for ventilation design were developed and used in many countries.

In 1851 Dr. John Gorrie was granted U.S. patent 8080 for a refrigeration machine. By the 1880s, refrigeration became available for industrial purposes. Initially, the two main uses were freezing meat for transport and making ice. However, in the early 1900s there was a new initiative to keep buildings cool for comfort. Cooling the New York Stock Exchange, in 1902, was one of the first comfort cooling systems. Comfort cooling was called "air conditioning."

Our title, "HVAC," thus captures the development of our industry. The term "air conditioning" has gradually changed, from meaning just cooling, to the total control of:

- Temperature
- Moisture in the air (humidity)
- Supply of outside air for ventilation
- Filtration of airborne particles
- Air movement in the occupied space

Throughout the rest of this text we will use the term "air conditioning" to include all of these issues and continue to use "HVAC" where only some of the elements of full air conditioning are being controlled.

To study the historical record of HVAC is to take a fascinating trip through the tremendous technical and scientific record of society. There are the pioneers such as Robert Boyle, Sadi Carnot, John Dalton, James Watt, Benjamin Franklin, John Gorrie, Lord Kelvin, Ferdinand Carré, Willis Carrier and Thomas Midgley, along with many others, who have brought us to our current state. Air-conditioning technology has developed since 1900 through the joint accomplishments of science and engineering. Advances in thermodynamics, fluid mechanics, electricity, electronics, construction, materials, medicine, controls and social behavior are the building blocks to better engineered products of air conditioning.

Historical accounts are not required as part of this course but, for the enjoyment and perspective it provides, it is worth reading an article such as "Milestones in Air Conditioning," by Walter A. Grant[1] or the book about Willis Carrier, *The Father of Air Conditioning*.[2] The textbook *Principles of Heating, Ventilating, and Air Conditioning*,[3] starts with a concise and comprehensive history of the HVAC industry.

HVAC evolved based on:

- Technological discoveries, such as refrigeration, that were quickly adopted for food storage.
- Economic pressures, such as the reduction in ventilation rates after the 1973 energy crisis.
- Computerization and networking, used for sophisticated control of large complex systems serving numerous buildings.
- Medical discoveries, such as the effects of second hand smoke on people, which influenced ventilation methods.

## 1.3 Scope of Modern HVAC

Modern air conditioning is critical to almost every facet of advancing human activity. Although there have been great advances in HVAC, there are several areas where active research and debate continue.

*Indoor air quality* is one that directly affects us. In many countries of the world there is a rapid rise in asthmatics and increasing dissatisfaction with indoor-air-quality in buildings and planes. The causes and effects are extremely complex. A significant scientific and engineering field has developed to investigate and address these issues.

*Greenhouse gas emissions* and the destruction of the earth's protective *ozone layer* are concerns that are stimulating research. New legislation and guidelines are evolving that encourage: recycling; the use of new forms of energy; less energy usage; and low polluting materials, particularly refrigerants. All these issues have a significant impact on building design, including HVAC systems and the design codes.

*Energy conservation* is an ongoing challenge to find novel ways to reduce consumption in new and existing buildings without compromising comfort and indoor air quality. Energy conservation requires significant cooperation between disciplines.

For example, electric lighting produces heat. When a system is in a cooling mode, this heat is an additional cooling load. Conversely, when the system is in a heating mode, the lighting heat reduces the load on the building heating system. This interaction between lighting and HVAC is the reason that ASHRAE and the Illuminating Engineering Society of North America (IESNA) joined forces to write the building energy conservation standard, *Standard 90.1–2004, Energy Standard for Buildings Except Low-Rise Residential Buildings*[4].

## 1.4 Introduction to Air-conditioning Processes

As mentioned earlier, the term "air conditioning," when properly used, now means the total control of temperature, moisture in the air (humidity), supply of outside air for ventilation, filtration of airborne particles, and air movement in the occupied space. There are seven main processes required to achieve full air conditioning and they are listed and explained below:

The processes are:

1. *Heating*—the process of adding thermal energy (heat) to the conditioned space for the purposes of raising or maintaining the temperature of the space.
2. *Cooling*—the process of removing thermal energy (heat) from the conditioned space for the purposes of lowering or maintaining the temperature of the space.
3. *Humidifying*—the process of adding water vapor (moisture) to the air in the conditioned space for the purposes of raising or maintaining the moisture content of the air.
4. *Dehumidifying*—the process of removing water vapor (moisture) from the air in the conditioned space for the purposes of lowering or maintaining the moisture content of the air.
5. *Cleaning*—the process of removing particulates, (dust etc.,) and biological contaminants, (insects, pollen etc.,) from the air delivered to the conditioned space for the purposes of improving or maintaining the air quality.

6. *Ventilating*—the process of exchanging air between the outdoors and the conditioned space for the purposes of diluting the gaseous contaminants in the air and improving or maintaining air quality, composition and freshness. Ventilation can be achieved either through *natural ventilation* or *mechanical ventilation*. Natural ventilation is driven by natural draft, like when you open a window. Mechanical ventilation can be achieved by using fans to draw air in from outside or by fans that exhaust air from the space to outside.
7. *Air Movement*—the process of circulating and mixing air through conditioned spaces in the building for the purposes of achieving the proper ventilation and facilitating the thermal energy transfer.

The requirements and importance of the seven processes varies. In a climate that stays warm all year, heating may not be required at all. Conversely, in a cold climate the periods of heat in the summer may be so infrequent as to make cooling unnecessary. In a dry desert climate, dehumidification may be redundant, and in a hot, humid climate dehumidification may be the most important design aspect of the air-conditioning system.

### Defining Air conditioning

The actual use of the words "air conditioning" varies considerably, so it is always advisable to check what is really meant. Consider, for example, "window air conditioners." The vast majority provide cooling, some dehumidification, some filtering, and some ventilation when the outside temperature is well above freezing. They have no ability to heat or to humidify the conditioned space and do not cool if it is cold outside.

In colder climates, heating is often provided by a separate, perimeter heating system, that is located within the outside walls. The other functions: cooling, humidification, dehumidification, cleaning, ventilating and air movement, are all provided by a separate air system, often referred to as the "air-conditioning system." **It is important to remember that both the heating and the air system together form the "air-conditioning" system for the space**.

## 1.5 Objective: What is your system to achieve?

Before starting to design a system, it is critical that you know what your system is to achieve.

Often, the objective is to provide a comfortable environment for the human occupants, but there are many other possible objectives: creating a suitable environment for farm animals; regulating a hospital operating room; maintaining cold temperatures for frozen food storage; or maintaining temperature and humidity to preserve wood and fiber works of art. Whatever the situation, it is important that the objective criteria for system success are clearly identified at the start of the project, because different requirements need different design considerations.

Let us very briefly consider some specific design situations and the types of performance requirements for HVAC systems.

**Example 1**: *Farm animals*. The design issues are economics, the health and well being of both animals and workers, plus any regulations. Farm animal spaces are always ventilated. Depending on the climate, cooling and/or heating may be provided, controlled by a simple thermostat. The ventilation rate may be varied to:

- Maintain indoor air quality (removal of body and excrement fumes.)
- Maintain inside design temperature (bring in cool air and exhaust hot air.)
- Remove moisture (bring in drier air and exhaust moist air.)
- Change the air movement over the animals (higher air speed provides cooling.)

A complex control of ventilation to meet the four design requirements may well be very cost effective. However, humidification and cleaning are not required.

**Example 2**: *Hospital operating room*. This is a critical environment, often served by a dedicated air-conditioning system. The design objectives include:

- Heating, to avoid the patient from becoming too cold.
- Cooling, to prevent the members of the operating team from becoming too hot.
- Control adjustment by the operating team for temperatures between 65°F (Fahrenheit) and 80°F.
- Humidifying, to avoid low humidity and the possibility of static electricity sparks.
- Dehumidifying, to minimize any possibility of mold and to minimize operating team discomfort.
- Cleaning the incoming air with very high efficiency filters, to remove any airborne organisms that could infect the patient.
- Ventilating, to remove airborne contaminants and to keep the theatre fresh.
- Providing steady air movement from ceiling supply air outlets down over the patient for exhaust near the floor, to minimize contamination of the operating site.

This situation requires a very comprehensive air-conditioning system.

**Example 3**: *Frozen food storage*. The ideal temperature for long storage varies: i.e. ice cream requires temperatures below −12°F and meat requires temperatures below −5°F. The design challenge is to ensure that the temperature is accurately maintained and that the temperature is as even as possible throughout the storage facility. Here, accurate cooling and good air movement are the prime issues. Although cooling and air movement are required, we refer to this system as a "freezer," not as an air-conditioning system, because heating, ventilation, humidification and dehumidification are not controlled.

**Example 4**: *Preserving wood and fiber works of art*. The objectives in this environment are to minimize any possibility of mold, by keeping the humidity low, and to minimize drying out, by keeping the humidity up. In addition, it is important to minimize the expansion and contraction of specimens that can

6    Fundamentals of HVAC

occur as the moisture content changes. As a result the design challenge is to maintain a very steady humidity, reasonably steady temperature, and to minimize required ventilation, from a system that runs continuously. For this situation, the humidity control is the primary issue and temperature control is secondary. Typically, this situation will require all seven of the air-conditioning features and we will describe the space as fully "air-conditioned."

Now let us go on to consider the more complex subject of human comfort in a space.

## 1.6 Environment For Human Comfort

"Provide a comfortable environment for the occupants" sounds like a simple objective, until you start to consider the variety of factors that influence the comfort of an individual. *Figure 1.1* is a simplified diagram of the three main groups of factors that affect comfort.

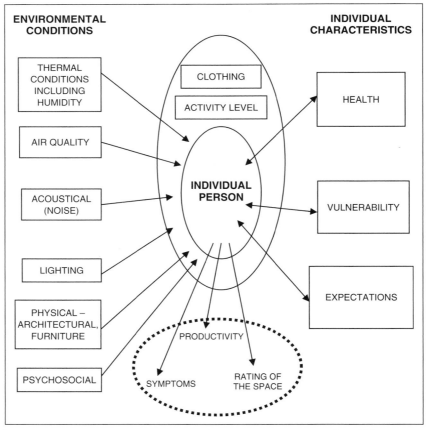

**Figure 1.1**   Personal Environment Model (adapted with permission from "The construct of comfort: a framework for research" by W.S. Cain[5])

Introduction to HVAC   7

- Attributes of the space – on the left
- Characteristics of the individual – on the right
- Clothing and activity of the individual – high center

### 1.6.1 Attributes of the Space Influencing Comfort

As you can see, six attributes of the space influence comfort: thermal, air quality, acoustical, lighting, physical, and psychosocial. Of these, only the thermal conditions and air quality can be directly controlled by the HVAC system. The acoustical (noise) environment may be influenced to some extent. The lighting and architectural aspects are another field, but these can influence how the HVAC is perceived. The psychosocial environment (how people interact sociably, or unsociably!) in the space is largely dependent on the occupants, rather than the design of the space.

We will briefly consider these six aspects of the space and their influence on comfort.

1. *Thermal conditions* include more than simply the air temperature. If the air speed is very high, the space will be considered drafty. If there is no air movement, occupants may consider the space 'stuffy'. The air velocity in a mechanically conditioned space is largely controlled by the design of the system.

    On the other hand, suppose the occupants are seated by a large unshaded window. If the air temperature stays constant, they will feel very warm when the sun is shining on them and cooler when clouds hide the sun. This is a situation where the architectural design of the space affects the thermal comfort of the occupant, independently of the temperature of the space.
2. The *air quality* in a space is affected by pollution from the occupants and other contents of the space. This pollution is, to a greater or lesser extent, reduced by the amount of outside air brought into the space to dilute the pollutants. Typically, densely occupied spaces, like movie theatres, and heavy polluting activities, such as cooking, require a much higher amount of outside air than an office building or a residence.
3. The *acoustical* environment may be affected by outside traffic noise, other occupants, equipment, and the HVAC system. Design requirements are dictated by the space. A designer may have to be very careful to design a virtually silent system for a recording studio. On the other hand, the design for a noisy foundry may not require any acoustical design consideration.
4. The *lighting* influences the HVAC design, since all lights give off heat. The lighting also influences the occupants' perception of comfort. If the lights are much too bright, the occupants may feel uncomfortable.
5. The *physical* aspects of the space that have an influence on the occupants include both the architectural design aspects of the space, and the interior design. Issues like chair comfort, the height of computer keyboards, or reflections off computer screens have no relation to the HVAC design, however they may affect how occupants perceive the overall comfort of the space.
6. The *psychosocial* situation, the interaction between people in the space, is not a design issue but can create strong feelings about the comfort of the space.

### 1.6.2 Characteristics of the Individual that Influence Comfort

Now let us consider the characteristics of the occupants of the space. All people bring with them health, vulnerabilities and expectations.

Their *health* may be excellent and they may not even notice the draft from the air conditioning. On the other hand if the occupants are patients in a doctor's waiting room, they could perceive a cold draft as very uncomfortable and distressing.

The occupants can also vary in *vulnerability*. For example, cool floors will likely not affect an active adult who is wearing shoes. The same floor may be uncomfortably cold for the baby who is crawling around on it.

Lastly the occupants bring their *expectations*. When we enter a prestigious hotel, we expect it to be comfortable. When we enter an air-conditioned building in summer, we expect it to be cool. The expectations may be based on previous experience in the space or based on the visual perception of the space. For example, when you enter the changing room in the gym, you expect it to be smelly, and your expectations make you more tolerant of the reality.

### 1.6.3 Clothing and Activity as a function of Individual Comfort

The third group of factors influencing comfort is the amount of clothing and the activity level of the individual. If we are wearing light clothing, the space needs to be warmer for comfort than if we are heavily clothed. Similarly, when we are involved in strenuous activity, we generate considerable body heat and are comfortable with a lower space temperature.

In the summer, in many business offices, managers wear suits with shirts and jackets while staff members may have bare arms, and light clothing. The same space may be thermally comfortable to one group and uncomfortable to the other.

There is much more to comfort than most people realize. These various aspects of comfort will be covered in more detail in later chapters.

## The Next Step

Chapter 2 introduces the concept of an air-conditioning system. We will then consider characteristics of systems and how various parameters influence system choice. Chapter 2 is broad in scope and will introduce you the content and value of the other ASHRAE Self-Study Courses.

## Summary

This has been an introduction to heating, ventilating and air conditioning and some of the terminology and main processes that are involved in air conditioning.

### 1.2 Brief History of HVAC

The field of HVAC started in the mid 1800s. The term "air conditioning" has gradually changed from meaning just cooling, to the total control of temperature, moisture in the air (humidity), supply of outside air for ventilation, filtration of airborne particles and air movement in the occupied space.

### 1.3 Scope of Modern HVAC

Some of the areas of research, regulation and responsibility include indoor air quality, greenhouse gas emissions, and energy conservation.

### 1.4 Introduction to Air-conditioning Processes

**There are seven main processes** required to achieve full air conditioning: heating, cooling, humidifying, dehumidifying, cleaning, ventilating, air movement. The requirements and importance of the seven processes vary with the climate.

### 1.5 System Objectives

Before starting to design a system, it is critical that you know what your system is supposed to achieve. The objective will determine the type of system to select, and the performance goals for it.

### 1.6 Environment For Human Comfort

The requirements for human comfort are affected by: the physical space; the characteristics of the individual, including health, vulnerability and expectations; and the clothing and activities of the individual.

Six attributes of the physical space that influence comfort are thermal, air quality, acoustical, lighting, physical, and the psychosocial environment. Of these, only the thermal conditions and air quality can be directly controlled by the HVAC system. The acoustical (noise) environment may be influenced to some extent. The lighting and architectural aspects can influence how the HVAC is perceived. The psychosocial environment in the space is largely dependent on the occupants rather than the design of the space.

## Bibliography

1. Grant, W. 1969. "Milestones in Air Conditioning." *ASHRAE Journal*. Atlanta: ASHRAE. Vol. 11, No. 9, pp. 45–51.
2. Ingels, M. 1991. *The Father of Air Conditioning*. Louisville, KY: Fetter Printing Co.
3. Sauer, Harry J. Jr., Ronald H. Howell, William J. Coad. 2001. *Principles of Heating, Ventilating, and Air Conditioning*. Atlanta: ASHRAE.
4. *Standard 90.1–2004 Energy Standard for Buildings Except Low-Rise Residential Buildings*. Atlanta: ASHRAE.
5. Cain, W.S. 2002. "The construct of comfort: a framework for research" Indoor Air 2002, *Proceedings: Indoor Air 2002* Volume II, pp.12–20.

# Chapter 2
# Introduction to HVAC Systems

## Contents of Chapter 2

Study Objectives of Chapter 2
2.1 Introduction
2.2 Introducing the Psychrometric Chart
2.3 Basic Air-Conditioning System
2.4 Zoned Air-Conditioning Systems
2.5 Choosing an Air-Conditioning System
2.6 System Choice Matrix
The Next Step
Summary
Bibliography

## Study Objectives of Chapter 2

Chapter 2 begins with an introduction to a graphical representation of air-conditioning processes called the psychrometric chart. Next, an air-conditioning system is introduced followed by a discussion about how it can be adapted to serve many spaces. The chapter ends with a brief introduction to the idea of using a factor matrix to help choose an air-conditioning system.

Chapter 2 is broad in scope and will also introduce you to the content and value of other, more in depth, ASHRAE Self-Study Courses. After studying Chapter 2, you should be able to:

Understand and describe the major concepts of the psychrometric chart.
Define the main issues to be considered when designing a system.
Name the four major system types and explain their differences.
Describe the main factors to be considered in a matrix selection process.

## 2.1 Introduction

In Chapter 1 we introduced the seven main air-conditioning processes and the task of establishing objectives for air-conditioning design. In this chapter we will consider

How these processes are described graphically in the psychrometric chart.
How these processes are combined to form an air-conditioning system.

The range of heating, ventilating and air-conditioning systems.
How system choices are made.

## 2.2 Introducing the Psychrometric Chart

Many of the air-conditioning processes involve air that is experiencing energy changes. These changes arise from changes in the air's temperature and its moisture content. The relationships between temperature, moisture content, and energy are most easily understood using a visual aid called the **"psychrometric chart."**

The psychrometric chart is an industry-standard tool that is used to visualize the interrelationships between dry air, moisture and energy. If you are responsible for the design or maintenance of any aspect of air conditioning in buildings, a clear and comfortable understanding of the chart will make your job easier.

Initially, the chart can be intimidating, but as you work with it you will discover that the relationships that it illustrates are relatively easy to understand. Once you are comfortable with it, you will discover that it is a tool that can make it easier to troubleshoot air-conditioning problems in buildings. The ASHRAE course, *Fundamentals of Thermodynamics and Psychrometrics*[1] goes into great detail about the use of the chart. That course also provides calculations and discussion about how the chart can be used as a design and troubleshooting tool.

In this course, however, we will only introduce the psychrometric chart, and provide a very brief overview of its structure.

### *The Design of the Psychrometric Chart*

The psychrometric chart is built upon two simple concepts.

1. Indoor air is a mixture of dry air and water vapor.
2. There is a specific amount of energy in the mixture at a specific temperature and pressure.

*Psychrometric Chart Concept 1: Indoor Air is a Mixture of Dry Air and Water Vapor.*

The air we live in is a mixture of both dry air and **water vapor**. Both are invisible gases. The water vapor in air is also called **moisture** or **humidity**. The quantity of water vapor in air is expressed as **"pounds of water vapor per pound of air."** This ratio is called the "humidity ratio," abbreviation W and the units are pounds of water/pound of dry air, $lb_w/lb_{da}$, often abbreviated to lb/lb.

The exact properties of moist air vary with pressure. Because pressure reduces as altitude increases, the properties of moist air change with altitude. Typically, psychrometric charts are printed based on standard pressure at sea level. For the rest of this course we will consider pressure as constant.

To understand the relationship between water vapor, air and temperature, we will consider two conditions:

**First Condition**: The temperature is constant, but the quantity of water vapor is increasing.

If the temperature remains constant, then, as the quantity of water vapor in the air increases, the humidity increases. However, at every temperature point, there is a maximum amount of water vapor that can co-exist with the air. The point at which this maximum is reached is called the **saturation point**. If more water vapor is added after the saturation point is reached, then an equal amount of water vapor condenses, and takes the form of either water droplets or ice crystals.

Outdoors, we see water droplets in the air as fog, clouds or rain and we see ice crystals in the air as snow or hail. The psychrometric chart only considers the conditions up to the saturation point; therefore, it only considers the effects of water in the vapor phase, and does not deal with water droplets or ice crystals.

**Second Condition**: The temperature is dropping, but the quantity of water vapor is constant.

If the air is cooled sufficiently, it reaches the **saturation line**. If it is cooled even more, moisture will condense out and dew forms.

For example, if a cold canned drink is taken out of the refrigerator and left for a few minutes, the container gets damp. This is because the moist air is in contact with the chilled container. The container cools the air that it contacts to a temperature that is below saturation, and dew forms. This temperature, at which the air starts to produce condensation, is called the **dew point temperature**.

### Relative Humidity

*Figure 2.1* is a plot of the maximum quantity of water vapor per pound of air against air temperature. The X-axis is temperature. The Y-axis is the proportion of water vapor to dry air, measured in pounds of water vapor per pound of dry air. The curved "maximum water vapor line" is called the "saturation line." It is also known as **100% relative humidity**, abbreviated to **100% rh**. At any point on the saturation line, the air has 100% of the water vapor per pound of air that can coexist with dry air at that temperature.

When the same volume of air contains only half the weight of water vapor that it has the capacity to hold at that temperature, we call it **50% relative humidity** or **50% rh**. This is shown in *Figure 2.2*. Air at any point on the 50% rh line has half the water vapor that the same volume of air could have at that temperature.

As you can see on the chart, the maximum amount of water vapor that moist air can contain increases rapidly with increasing temperature. For example, moist air at the freezing point, 32°F, can contain only 0.4% of its weight as

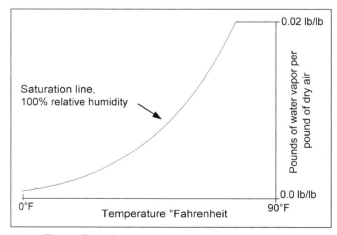

**Figure 2.1** Psychrometric Chart – Saturation Line

Introduction to HVAC Systems    13

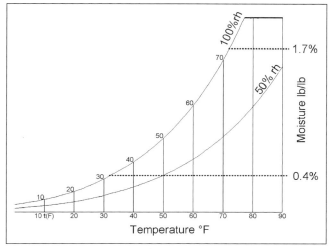

**Figure 2.2**  Psychrometric Chart – 50% Relative Humidity Line

water vapor. However, indoors, at a temperature of 72°F the moist air can contain nearly 1.7% of its weight as water vapor—over four times as much.

Consider *Figure 2.3*, and this example:

On a miserable wet day it might be 36°F outside, with the air rather humid, at 70% relative humidity. Bring that air into your building. Heat it to 70°F. This brings the relative humidity down to about 20%. This change in relative humidity is shown in *Figure 2.3*, from **Point 1 → 2**. A cool damp day outside provides air for a dry day indoors! Note that the absolute amount of water vapor in the air has remained the same, at 0.003 pounds of water vapor per pound of dry air; but as the temperature rises, the relative humidity falls.

Here is an example for you to try, using *Figure 2.3*.

Suppose it is a warm day with an outside temperature of 90°F and relative humidity at 40%. We have an air-conditioned space that is at 73°F. Some

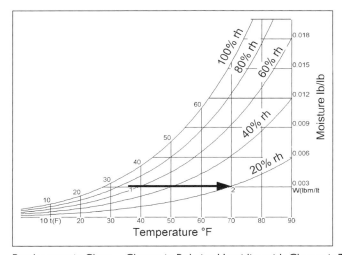

**Figure 2.3**  Psychrometric Chart – Change in Relative Humidity with Change in Temperature

of the outside air leaks into our air-conditioned space. This leakage is called **infiltration**.

Plot the process on *Figure 2.3*.
Find the start condition, 90°F and 40% rh, moisture content 0.012 lb/lb.
Then cool this air: move left, at constant moisture content to 73°F.
Notice that the cooled air now has a relative humidity of about 70%.

Relative humidity of 70% is high enough to cause mold problems in buildings. Therefore in hot moist climates, to prevent infiltration and mold generation, it is valuable to maintain a small positive pressure in buildings.

*Psychrometric Chart Concept 2: There is a specific amount of energy in the air mixture at a specific temperature and pressure.*

This brings us to the second concept that the psychrometric chart illustrates. There is a specific amount of energy in the air water-vapor mixture at a specific temperature. The energy of this mixture is dependent on two measures:

1. The temperature of the air.
2. The proportion of water vapor in the air.

There is more energy in air at higher temperatures. The addition of heat to raise the temperature is called adding "**sensible heat**." There is also more energy when there is more water vapor in the air. The energy that the water vapor contains is referred to as its "**latent heat**."

The measure of the total energy of both the sensible heat in the air and the latent heat in the water vapor is commonly called "**enthalpy**." Enthalpy can be raised by adding energy to the mixture of dry air and water vapor. This can be accomplished by adding either or both

- Sensible heat to the air
- More water vapor, which increases the latent heat of the mixture

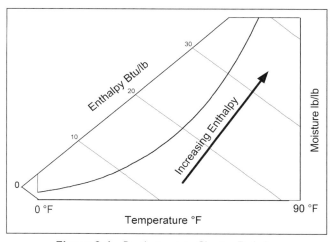

**Figure 2.4** Psychrometric Chart – Enthalpy

On the psychrometric chart, lines of constant enthalpy slope down from left to right as shown in *Figure 2.4* and are labeled "Enthalpy."

The zero is arbitrarily chosen as zero at 0°F and zero moisture content. The unit measure for enthalpy is **British Thermal Units per pound of dry air**, abbreviated as **Btu/lb**.

## Heating

The process of heating involves the addition of sensible heat energy. *Figure 2.5* illustrates outside air at 47°F and almost 90% relative humidity that has been heated to 72°F. This process increases the enthalpy in the air from approximately 18 Btu/lb to 24 Btu/lb. Note that the process line is **horizontal** because no water vapor is being added to, or removed from the air—we are just heating the mixture. In the process, the relative humidity drops from almost 90% rh down to about 36% rh.

Here is an example for you to try.

Plot this process on Figure 2.6.

Suppose it is a cool day with an outside temperature of 40°F and 60% rh. We have an air-conditioned space and the air is heated to 70°F. There is no change in the amount of water vapor in the air. The enthalpy rises from about 13 Btu/lb to 20 Btu/lb, an increase of 7 Btu/lb.

As you can see, the humidity would have dropped to 20% rh. This is quite dry so let us assume that we are to raise the humidity to a more comfortable 40%. As you can see on the chart, this raises the enthalpy by an additional 3.5 Btu/lb.

## Humidification

The addition of water vapor to air is a process called "**humidification**." Humidification occurs when water absorbs energy, evaporates into water vapor, and mixes with air. The energy that the water absorbs is called "**latent heat**."

There are two ways for humidification to occur. In both methods, energy is added to the water to create water vapor.

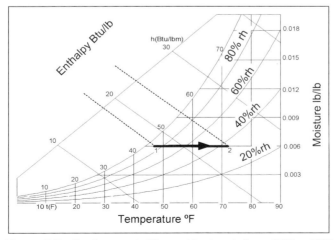

**Figure 2.5** Psychrometric Chart – Heating air from 47°F to 72°F

16    Fundamentals of HVAC

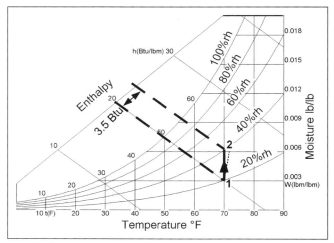

**Figure 2.6**  Psychrometric Chart – Adding Moisture with Steam

1. **Water can be heated**. When heat energy is added to the water, the water is transformed to its gaseous state, steam, that mixes into the air. In *Figure 2.6*, the vertical line, from Point 1 to Point 2, shows this process. The heat energy, 3.5 Btu/lb, is put into the water to generate steam (vaporize it), which is then mixed with the air.

    In practical steam humidifiers, the added steam is hotter than the air and the piping loses some heat into the air. Therefore, the air is both humidified and heated due to the addition of the water vapor. This combined humidification and heating is shown by the dotted line which slopes a little to the right in *Figure 2.6*.

2. **Water can evaporate** by spraying a fine mist of water droplets into the air. The fine water droplets absorb heat from the air as they evaporate. Alternatively, but using the same evaporation process, air can be passed over a wet fabric, or wet surface, enabling the water to evaporate into the air.

    In an evaporative humidifier, the evaporating water absorbs heat from the air to provide its latent heat for evaporation. As a result, the air temperature drops as it is humidified. The process occurs with no external addition or removal of heat. It is called an **adiabatic process**. Since there is no change in the heat energy (enthalpy) in the air stream, the addition of moisture, by evaporation, occurs along a line of constant enthalpy.

    *Figure 2.7* shows the process. From Point 1, the moisture evaporates into the air and the temperature falls to 56°F, Point 2. During this evaporation, the relative humidity rises to about 65%. To reach our target of 70°F and 40% rh we must now heat the moistened air at Point 2 from 56°F to 70°F, Point 3, requiring 3.5 Btu/lb of dry air.

To summarize, we can humidify by adding heat to water to produce steam and mixing the steam with the air, or we can evaporate the moisture and heat the moistened air. We achieve the same result with the same input of heat by two different methods.

The process of evaporative cooling can be used very effectively in a hot, dry desert climate to pre-cool the incoming ventilation air. For example,

Introduction to HVAC Systems    17

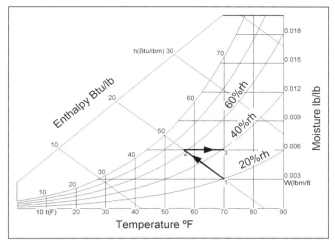

**Figure 2.7**   Psychrometric Chart – Adding Moisture, Evaporative Humidifier

outside air at 90°F and 15% relative humidity could be cooled to 82°F by passing it through an evaporative cooler. The relative humidity will rise, but only to about 27%. Even with no mechanical refrigeration, this results in a pleasant reduction in air temperature without raising the relative humidity excessively.

### Cooling and Dehumidification

Cooling is most often achieved in an air-conditioning system by passing the moist air over a cooling coil. As illustrated in *Figure 2.8*, a coil is constructed of a long serpentine pipe through which a cold liquid or gas flows. This cold fluid is either chilled water, typically between 40°F and 45°F, or a refrigerant. The pipe is lined with fins to increase the heat transfer from the air to the cold fluid in the pipe. *Figure 2.8* shows the face of the coil, in the direction of airflow. Depending on the coil design, required temperature

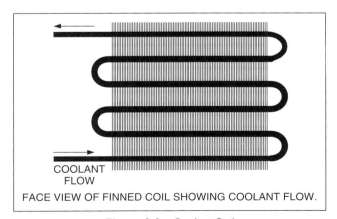

**Figure 2.8**   Cooling Coil

18  Fundamentals of HVAC

drop, and moisture removal performance, the coil may have 2 to 8 rows of piping. Generally the more rows, the higher the moisture removal ability of the coil.

There are two results. First, the cooling coil cools the air as the air passes over the coils. Second, because the cooling fluid in the coil is usually well below the saturation temperature of the air, moisture condenses on the coil, and drips off, to drain away. This process reduces the enthalpy, or heat, of the air mixture and increases the enthalpy of the chilled water or refrigerant. In another part of the system, this added heat must be removed from the chilled water or refrigerant to recool it for reuse in the cooling coil.

The amount of moisture that is removed depends on several factors including:

- The temperature of the cooling fluid
- The depth of the coil
- Whether the fins are flat or embossed
- The air velocity across the coil.

An example of the typical process is shown in *Figure 2.9*.

The warm moist air comes into the building at 80°F and 50% rh, and passes through a cooling coil. In this process, the air is being cooled to 57°F. As the moisture condenses on the coil, it releases its latent heat and this heat has to be removed by the cooling fluid. In *Figure 2.9* the moisture removal enthalpy, A → B, is about a third of the enthalpy required to cool the air, B → C.

This has been a very brief introduction to the concepts of the psychrometric chart. A typical chart is shown in *Figure 2.10*. It looks complicated, but you know the simple underlying ideas:

Indoor air is a mixture of dry air and water vapor.
There is a specific amount of total energy, called enthalpy, in the mixture at a specific temperature, moisture content and pressure.
There is a maximum limit to the amount of water vapor in the mixture at any particular temperature.

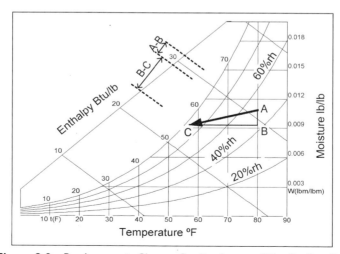

**Figure 2.9**  Psychrometric Chart – Cooling Across a Wet Cooling Coil

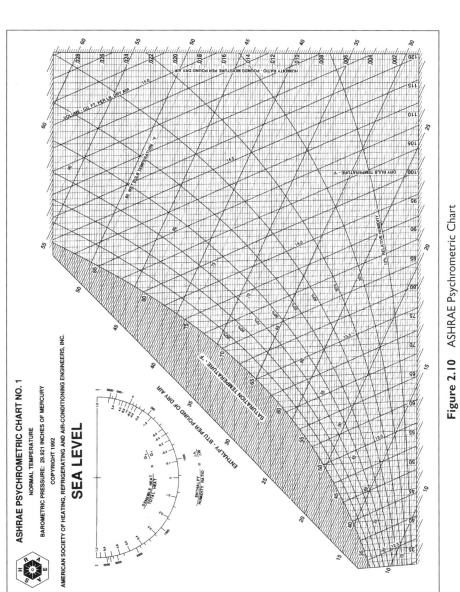

**Figure 2.10** ASHRAE Psychrometric Chart

20    Fundamentals of HVAC

The actual use of the chart for design, including the calculations, is detailed in the ASHRAE course *Fundamentals of Thermodynamics and Psychrometrics*[1].

Now that we have an understanding of the relationships of dry air, moisture and energy, at a particular pressure we will consider an air-conditioning plant that will provide all seven basic functions of an air-conditioning system to a single space. Remember, the processes required are: heating, cooling, dehumidifying, humidifying, ventilating, cleaning and air movement.

## 2.3 Basic Air-Conditioning System

*Figure 2.11* shows the schematic diagram of an air-conditioning plant. The majority of the air is drawn from the space, mixed with outside ventilation air and then conditioned before being blown back into the space.

As you discovered in Chapter 1, air-conditioning systems are designed to meet a variety of objectives. In many commercial and institutional systems, the ratio of outside ventilation air to return air typically varies from 15 to 25% of outside air. There are, however, systems which provide 100% outside air with zero recirculation.

The components, from left to right, are:

**Outside Air Damper**, which closes off the outside air intake when the system is switched off. The damper can be on a spring return with a motor to drive it open; then it will automatically close on power failure. On many systems there will be a metal mesh screen located upstream of the filter, to prevent birds and small animals from entering, and to catch larger items such as leaves and pieces of paper.

**Mixing chamber**, where return air from the space is mixed with the outside ventilation air.

**Filter**, which cleans the air by removing solid airborne contaminants (dirt). The filter is positioned so that it cleans the return air and the ventilation air. The filter is also positioned upstream of any heating or cooling coils, to keep the coils clean. This is particularly important for the cooling coil, because the coil is wet with condensation when it is cooling.

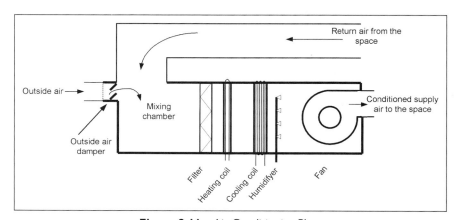

**Figure 2.11**    Air-Conditioning Plant

**Heating coil**, which raises the air temperature to the required supply temperature.

**Cooling coil**, which provides cooling and dehumidification. A thermostat mounted in the space will normally control this coil. A single thermostat and controller are often used to control both the heating and cooling coil. This method reduces energy waste, because it ensures the two coils cannot both be "on" at the same time.

**Humidifier**, which adds moisture, and which is usually controlled by a **humidistat** in the space. In addition, a high humidity override humidistat will often be mounted just downstream of the fan, to switch the humidification "off" if it is too humid in the duct. This minimizes the possibility of condensation forming in the duct.

**Fan**, to draw the air through the resistance of the system and blow it into the space.

These components are controlled to achieve six of the seven air-conditioning processes.

*Heating*: directly by the space thermostat controlling the amount of heat supplied by the heating coil.

*Cooling*: directly by the space thermostat controlling the amount of cooling supplied to the cooling coil.

*Dehumidifying*: by default when cooling is required, since, as the cooling coil cools the air, some moisture condenses out.

*Humidifying*: directly, by releasing steam into the air, or by a very fine water spray into the air causing both humidification and cooling.

*Ventilating*: provided by the outside air brought in to the system.

*Cleaning*: provided by the supply of filtered air.

*Air movement* within the space is not addressed by the air-conditioning plant, but rather by the way the air is delivered into the space.

## *Economizer Cycle*

In many climates there are substantial periods of time when cooling is required and the return air from the space is warmer and moister than the outside air. During these periods, you can reduce the cooling load on the cooling coil by bringing in more outside air than that required for ventilation. This can be accomplished by expanding the design of the basic air-conditioning system to include an **economizer**.

The economizer consists of three (or four) additional components as shown in *Figure 2.12*.

**Expanded air intake and damper**, sized for 100% system flow.

**Relief air outlet with automatic damper**, to exhaust excess air to outside.

**Return air damper**, to adjust the flow of return air into the mixing chamber.

**(Optional) Return fan in the return air duct**. The return fan is often added on economizer systems, particularly on larger systems. If there is no return fan, the main supply fan must provide enough positive pressure in the space to force the return air out through any ducting and the relief dampers. This can cause unacceptable pressures in the space, making doors slam and difficult to open. When the return air fan is added it will overcome the resistance of the return duct and relief damper, so the space pressure stays near neutral to outside.

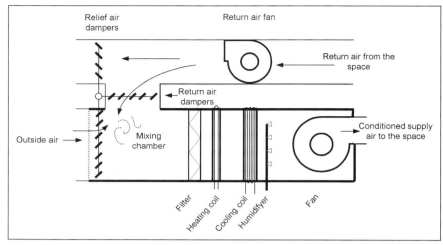

**Figure 2.12**  Air-Conditioning Plant with Economizer Cycle

Example: Let us consider the operation of the economizer system in *Figure 2.13*. The particular system operating requirements and settings are:

The system is required to provide supply air at 55°F
Return air from the space is at 75°F
Minimum outside air requirement is 20%,
Above 68°F, the system will revert to minimum outside air for ventilation.

In Figure 2.13, the outside temperature is shown along the x-axis from −60°F to +100°F. We are going to consider the economizer operation from −50°F up to 100°F, working across Figure 2.13 from left to right.

At −50°F, the minimum 20% outside air for ventilation is mixing with 80% return air at 75°F and will produce a mixed temperature of only 50°F. Therefore, in order to achieve the required supply air at 55°F, the heater will have to increase the temperature by 5°F.

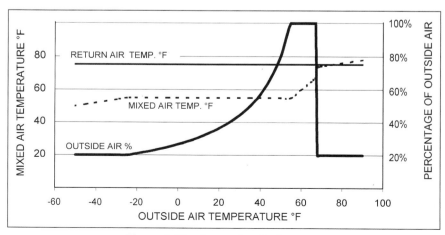

**Figure 2.13**  Economizer Performance

At −25°F, the minimum outside air for ventilation, 20%, is mixing with 80% return air at 75°F to produce a mixed temperature of 55°F, so the supply air will no longer require any additional heating.

As the temperature rises above −25°F the proportion of outside air will steadily increase to maintain a mixed temperature of 55°F. When the outside air temperature reaches 55°F the mixture will be 100% outside air (and 0% return air). This represents full economizer operation.

Above 55°F the controls will maintain 100% outside air but the temperature will rise as does the outside temperature. The cooling coil will come on to cool the mixed air to the required 55°F.

In this example, at 68°F the controls will close the outside air dampers, and allow only the required 20% ventilation air into the mixing chamber.

From 68°F to 100°F the system will be mixing 20% outside air and 80% return air. This will produce a mixture with temperature rising from 73.6°F to 80°F as the outside air temperature rises from 68°F to 100°F.

The useful economizer operation is from −25°F to 68°F. Below −25°F the economizer has no effect, since the system is operating with the minimum 20% outside ventilation air intake. In this example, 68°F was a predetermined change-over point. Above 68°F, the economizer turns off, and the system reverts to the minimum outside air amount, 20%.

The economizer is a very valuable energy saver for climates with long periods of cool weather. For climates with warm moist weather most of the year, the additional cost is not recovered in savings. Also, for spaces where the relative humidity must be maintained above ~45%, operation in very cold weather is uneconomic. This is because cold outside air is very dry, and considerable supplementary humidification energy is required to humidify the additional outside air.

## 2.4 Zoned Air-Conditioning Systems

The air-conditioning system considered so far provides a single source of air with uniform temperature to the entire space, controlled by one space thermostat and one space humidistat. However, in many buildings there is a variety of spaces with different users and varying thermal loads. These varying loads may be due to different inside uses of the spaces, or due to changes in cooling loads because the sun shines into some spaces and not others. Thus our simple system, which supplies a single source of heating or cooling, must be modified to provide independent, variable cooling or heating to each space.

When a system is designed to provide independent control in different spaces, each space is called a "**zone**." A zone may be a separate room. A zone may also be part of a large space. For example, a theatre stage may be a zone, while the audience seating area is a second zone in the same big space. Each has a different requirement for heating and cooling.

This need for zoning leads us to the four broad categories of air-conditioning systems, and consideration of how each can provide zoned cooling and heating. The four systems are

1. All-air systems
2. Air-and-water systems
3. All-water systems
4. Unitary, refrigeration-based systems

### System 1: All-air Systems

All-air systems provide air conditioning by using a tempered flow of air to the spaces. These all-air systems need substantial space for ducting the air to each zone.

The cooling or heating capacity, **Q**, is measured in British Thermal Units (Btu) and is the product of airflow, measured in cubic feet per minute, (**cfm**), times the difference in temperature between the supply air to the zone and the return air from the zone.

$$Q \text{ (Btu)} = \text{Constant} \cdot \text{mass flow} \cdot \text{temperature difference}$$

$$Q \text{ (Btu)} = \text{Constant} \cdot \text{cfm} \cdot (°F_{zone} - °F_{supply\ air})$$

To change the heating or cooling capacity of the air supply to one zone, the system must either alter the supply temperature, °F, or alter the flow, cfm, to that zone.

**Reheat system**: The simplest, and least energy efficient system, is the constant volume reheat system. Let us assume that the main air system provides air that is cool enough to satisfy all possible cooling loads, and that there is a heater in the duct to each zone.

A zone thermostat can then control the heater to maintain the desired zone set-point-temperature. The system, shown in *Figure 2.14*, is called a **reheat system**, since the cool air is reheated as necessary to maintain zone temperature.

*Figure 2.14* illustrates the basic air-conditioning system, plus ducting, to only two of many zones. The air to each zone passes over a reheat coil before entering the zone. A thermostat in the zone controls the reheat coil. If the zone requires full cooling, the thermostat will shut off the reheat coil. Then, as the cooling load drops, the thermostat will turn on the coil to maintain the zone temperature.

**Variable Air Volume (VAV) System**: *Figure 2.15* illustrates another zoned system, called a Variable Air Volume system, VAV system, because it varies the volume of air supplied to each zone.

Variable Air Volume systems are more energy efficient than the reheat systems. Again, assume that the basic system provides air that is cool enough to satisfy all possible cooling loads. In zones that require only cooling, the duct

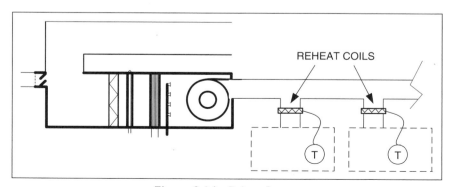

**Figure 2.14**  Reheat System

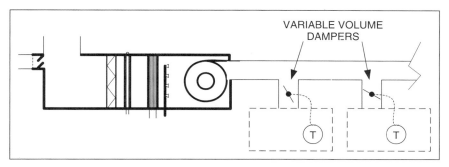

**Figure 2.15**   Variable Air Volume (VAV) System

to each zone can be fitted with a control damper that can be throttled to reduce the airflow to maintain the desired temperature.

In both types of systems, all the air-conditioning processes are achieved through the flow of air from a central unit into each zone. Therefore they are called "**all-air systems**." We will discuss these systems in a bit more detail in Chapter 7. However, to design and choose systems, you will need the detailed information found in the ASHRAE course *Fundamentals of Air System Design*[2].

### System 2: Air-and-water Systems

Another group of systems, air-and-water systems, provide all the primary ventilation air from a central system, but local units provide additional conditioning. The primary ventilation system also provides most, or all, of the humidity control by conditioning the ventilation air. The local units are usually supplied with hot or chilled water. These systems are particularly effective in perimeter spaces, where high heating and cooling loads occur. Although they may use electric coils instead of water, they are grouped under the title "**air-and-water systems**." For example, in cold climates substantial heating is often required at the perimeter walls. In this situation, a hot-water-heating system can be installed around the perimeter of the building while a central air system provides cooling and ventilation.

### System 3: All-water Systems

When the ventilation is provided through natural ventilation, by opening windows, or other means, there is no need to duct ventilation air to the zones from a central plant. This allows all processes other than ventilation to be provided by local equipment supplied with hot and chilled water from a central plant. These systems are grouped under the name "**all-water systems**."

The largest group of all-water systems are heating systems. We will introduce these systems, pumps and piping in Chapters 8 and 9. The detailed design of these heating systems is covered in the ASHRAE course *Fundamentals of Heating Systems*[3].

Both the air-and-water and all-water systems rely on a central supply of hot water for heating and chilled water for cooling. The detailed designs and calculations for these systems can be found in the ASHRAE course *Fundamentals of Water System Design*[4].

### System 4: Unitary, Refrigerant-based Systems

The final type of system uses local refrigeration equipment and heaters to provide air conditioning. They are called "**unitary refrigerant–based systems**" and we will discuss them in more detail in Chapter 6.

The window air-conditioner is the simplest example of this type of system. In these systems, ventilation air may be brought in by the unit, by opening windows, or from a central ventilation air system.

The unitary system has local refrigerant-based cooling. In comparison, the other types of systems use a central refrigeration unit to either cool the air-conditioning airflow or to chill water for circulation to local cooling units.

The design, operation and choice of refrigeration equipment is a huge field of knowledge in itself. Refrigeration equipment choices, design, installation, and operating issues are introduced in the ASHRAE course *Fundamentals of Refrigeration*[5].

### System Control

We have not yet considered how any of these systems can be controlled. Controls have become a vast area of knowledge with the use of solid-state sensors, computers, radio and the Internet. Basic concepts will be introduced throughout this text, with a focused discussion in Chapter 11. For an in-depth introduction to controls, ASHRAE provides the course *Fundamentals of HVAC Control Systems*[6].

## 2.5 Choosing an Air-Conditioning System

Each of the four general types of air-conditioning systems has numerous variations, so choosing a system is not a simple task. With experience, it becomes easier. However, a new client, a new type of building or a very different climate can be a challenge.

We are now going to briefly outline the range of factors that affect system choice and finish by introducing a process that designers can use to help choose a system.

The factors, or parameters that influence system choice can conveniently be divided into the following groups:

- Building design
- Location issues
- Utilities: availability and cost
- Indoor requirements and loads
- Client issues

### Building Design

The design of the building has a major influence on system choice. For example, if there is very little space for running ducts around the building, an all-air system may not fit in the available space.

### Location Issues

The building location determines the weather conditions that will affect the building and its occupants. For the specific location we will need to consider factors like:

site conditions
peak summer cooling conditions
summer humidity
peak winter heating conditions
wind speeds
sunshine hours
typical snow accumulation depths

The building location and, at times, the client, will determine what national, local, and facility specific codes must be followed. Typically, the designer must follow the local codes. These include:

*Building code* that includes a section on HVAC design including ventilation.
*Fire code* that specifies how the system must be designed to minimize the start and spread of fire and smoke.
*Energy code* that mandates minimum energy efficiencies for the building and components. We will be considering the ASHRAE Standard 90.1 2004 *Energy Standard for Buildings Except Low-Rise Residential Buildings*[7] and other energy conservation issues in Chapter 12.

In addition, some types of buildings, such as medical facilities, are designed to consensus codes which may not be required by local authorities but which may be mandated by the client. An example is The American Institute of Architects *Guidelines for Design and Construction of Hospital and Health Care Facilities*[8], which has guidelines that are extremely onerous in some climates.

### Utilities: Availability and Cost

The choice of system can be heavily influenced by available utilities and their costs to supply and use. So, if chilled water is available from the adjacent building, it would probably be cost advantageous to use it, rather than install new unitary refrigerant-based units in the new building.

Then again, the cost of electricity may be very high at peak periods, encouraging the design of an electrically-efficient system with low peak-demand for electricity. We will be introducing some of the ways to limit the cost of peak-time electricity in our final chapter, Chapter 13.

The issues around electrical pricing and usage have become very well publicized in North America over recent years. The ASHRAE course *Fundamentals of Electrical Systems and Building Electrical Energy Use*[10] introduces this topic.

### Indoor Requirements and Loads

The location effects and indoor requirements provide all the necessary information for load calculation for the systems.

**The thermal and moisture loads** – Occupants' requirements and heat output from lighting and equipment affect the demands on the air-conditioning system.
**Outside ventilation air** – The occupants and other polluting sources, such as cooking, will determine the requirements.
**Zoning** – The indoor arrangement of spaces and uses will determine if, and how, the system is to be zoned.

Other indoor restrictions may be very project, or even zone specific. For example, a sound recording studio requires an extremely quiet system and negligible vibration.

The methods of calculating the heating and cooling loads are fully explained, with examples, in the ASHRAE course *Fundamentals of Heating and Cooling Loads*[9].

### Client Issues

Buildings cost money to construct and to use. Therefore, the designer has to consider the clients' requirements both for construction and for in-use costs. For example, the available construction finances may dictate a very simple system. Alternatively, the client may wish to finance a very sophisticated, and more expensive system to achieve superior performance, or to reduce in-use costs.

In addition to cost structures, the availability of maintenance staff must be considered. A building at a very remote site should have simple, reliable systems, unless very competent and well-supported maintenance staff will be available.

Clients' approvals may be gained, or lost, based on their own previous experience with other projects or systems. Therefore, it is important for the designer to find out, in advance, if the client has existing preconceptions about potential systems.

### System Choice

While all the above factors are considered when choosing a system, the first step in making a choice is to calculate the system loads and establish the number and size of the zones. Understanding of the loads may eliminate some systems from consideration. For example:

- In warm climates where heating is not required only systems providing cooling need be considered.
- If there are significant variations in operating hours between zones, a system which cannot be shut down on a zone-by-zone basis may not be worth considering.

Typically, after some systems have been eliminated for specific reasons, one needs to do a point-by-point comparison to make a final choice. This is where the system-choice matrix is a very useful tool.

## 2.6 System Choice Matrix

The matrix method of system choice consists of a list of relevant factors that affect system choice and a tabular method of comparing the systems under consideration.

*Figure 2.16* provides an illustration of the matrix method of choosing a system. In the left column of the matrix are the relevant factors that will be used to evaluate the systems, and the top row shows the systems under consideration.

In our example, we have simplified the matrix in both dimensions. We have strictly limited our relevant factors, and we have limited our choices down to two systems, the reheat system and a VAV system. Note that in a real matrix you would include all the relevant issues, as discussed in the preceding section. You would also probably have several systems under consideration.

|  | Relative Importance | System 1 Reheat | | System 2 Variable Air Volume | |
|---|---|---|---|---|---|
|  |  | Relative Performance | Relative Score | Relative Performance | Relative Score |
| Cooling Capacity | 8 | 10 | 80 | 10 | 80 |
| Temperature Control | 9 | 10 | 90 | 8 | 72 |
| Zone Occupancy Timing | 10 | 1 | 10 | 9 | 90 |
| First Cost | 5 | 7 | 35 | 5 | 25 |
| Operating Cost | 8 | 3 | 24 | 8 | 64 |
|  |  | Totals | 239 |  | 331 |

**Figure 2.16** Matrix for Systems Choice

In this example, the relevant design issues for this building are as follows:

- The building requires cooling but no heating.
- Some areas of the building will be in use for 24 hours every day of the week. Other areas will be used just during the day, Monday to Friday.
- The client has indicated that operational expenses (ongoing) are more important than construction costs (one time).

As you can see, the matrix has a list of relevant issues down the left hand side. Each issue may have a greater or lesser importance. In the column headed "Relative importance" one assigns a multiplier between 1 and 10, with 10 meaning "extremely important" and 1 meaning "not important." So if, for our example, temperature control is very important it might be rated "9" and the ability to Zone – which is critical to economic operation in this particular building, requires a relative importance of 10. As you can see in the matrix, it is possible for two factors to share the same relative importance.

Once the relative importances have been assigned, it is time to assess the systems under consideration. In our example, both systems have excellent cooling capacity. They each score "10" under performance for this factor.

When we consider the requirement for zone occupancy-timing, however, we note that the reheat system does not have any ability to shut off one part of the system and leave another running. Therefore, it scores only "1" for this requirement. The VAV system, on the other hand, has the capacity to shut off any zone at any time though the main fan still has to run, even if only one zone is on. Therefore the VAV system scores "9" for this factor.

The VAV system also gets a higher score for first cost (construction cost) and for operating expense.

After each factor has been considered, the "relative performance" number is multiplied by the "relative importance" multiplier, to obtain the relative score for that item. The results for each system are totaled, and compared.

In this example, the VAV has a higher score and would be chosen.

The method is an excellent way of methodically assessing system alternatives. However, it should be used intelligently. If a system fails on a critical requirement, it should be eliminated, even if its total score may be the highest. For example, on a prison project, one would likely exclude any system that requires maintenance from the cells, regardless of how high it scored on a matrix!

For a more complete listing of issues for use in a matrix see Chapter 1 of ASHRAE *Systems and Equipment Handbook*[11] 2004 and for information on operating and other costs see Chapter 35 in the ASHRAE *Applications Handbook*[12] 2003.

## The Next Step

Having introduced systems and the range of design issues, the next two chapters will cover two specific subjects which dictate design requirements: **Thermal Comfort** in Chapter 3, and **Ventilation** and **Indoor Air Quality** in Chapter 4.

## Summary

### 2.2 The Psychrometric Chart

The psychrometric chart is a visual aid that demonstrates the relationships of air temperature, moisture content, and energy. It is built upon three simple concepts:

Indoor air is a mixture of dry air and water vapor.

At any given temperature, there is a maximum amount of water vapor that the mixture can sustain. The saturation line represents this maximum. When moist air is cooled to a temperature below the saturation line, the water vapor condenses, and the air is dehumidified. The addition of water to air is called humidification. This occurs when water absorbs energy, evaporates into water vapor and mixes with air. Humidification can take place when water is heated, to produce steam that mixes into the air, or when water evaporates into the air. Evaporation occurs with no external addition or removal of heat. It is called an "adiabatic process." The energy that the water vapor absorbs as it evaporates is referred to as its "latent heat."

There is a specific amount of energy in the dry air/water vapor mixture at a specific temperature and pressure. The energy of this mixture, at a particular pressure, is dependent on two measures: the temperature of the air, and the quantity of water vapor in the air. The total energy of the air/water vapor mixture is called "Enthalpy." The unit measure for enthalpy is British Thermal Units per pound of dry air, abbreviated as Btu/lb.

### 2.3 The Components of a Basic Air-Conditioning System

These include the outside air damper, the mixing chamber, the filter, the heating coil, the cooling coil, the humidifier and the fan. These components are controlled to achieve six of the seven air-conditioning processes: heating, humidifying, cooling, dehumidifying, ventilating, and cleaning.

The economizer cycle is an energy saver for climates with long periods of cool weather. The economizer consists of three, or four additional components: expanded air intake and damper sized for 100% flow; relief outlet with damper to exhaust excess air to outside; return air damper to adjust the flow of return air into the mixing chamber; (optional) return fan in the return air duct.

## 2.4 Zoned Air-Conditioning Systems

Zoning is used to provide variable heating or cooling in different spaces using: all-air systems, like reheat and variable air volume systems; air-and-water systems, all-water systems, and unitary, refrigeration-based systems.

## 2.5 Choosing an Air-Conditioning System

Design factors for choosing an air-conditioning system include: building design, location issues, utilities – availability and cost, indoor requirements and loads, and client issues.

## 2.6 System Choice Matrix

To determine the relative importance of the different design factors, you can use a System Choice Matrix to compare the systems that are under consideration.

## Bibliography

1. *Fundamentals of Thermodynamics and Psychrometrics*. ASHRAE Learning Institute Publication.
2. *Fundamentals of Air System Design*. ASHRAE Learning Institute Publication.
3. *Fundamentals of Heating Systems*. ASHRAE Learning Institute Publication.
4. *Fundamentals of Water System Design*. ASHRAE Learning Institute Publication.
5. *Fundamentals of Refrigeration*. ASHRAE Learning Institute Publication.
6. *Fundamentals of HVAC Control Systems*. ASHRAE Learning Institute Publication.
7. *Standard 90.1-2004 Energy Standard for Buildings Except Low-Rise Residential Buildings*. ASHRAE
8. *Guidelines for Design and Construction of Hospital and Health Care Facilities*. 2001. The American Institute of Architects Washington, DC
9. *Fundamentals of Heating and Cooling Loads*. ASHRAE Learning Institute Publication.
10. *Fundamentals of Electrical Systems and Building Electrical Energy Use*. ASHRAE Learning Institute Publication.
11. ASHRAE Systems and Equipment Handbook, 2004
12. ASHRAE Applications Handbook, 2003

Chapter 3

# Thermal Comfort

## Contents of Chapter 3

Study Objectives of Chapter 3
3.1 Introduction – What is Thermal Comfort?
3.2 Seven Factors Influencing Thermal Comfort
3.3 Conditions for Comfort
3.4 Managing Under Less Than Ideal Conditions
3.5 Requirements of Non-Standard Groups
The Next Step
Summary
Bibliography

## Study Objectives of Chapter 3

Having studied this chapter you should be able to:

List seven factors influencing thermal comfort.
Explain why thermal comfort depends on the individual as well as the thermal conditions.
Choose acceptable thermal design conditions.

## 3.1 Introduction: What is Thermal Comfort?

In Chapter 1, Sections 1.6.1 and 1.6.2, we introduced the Personal Environmental Model that illustrated the main factors that affect human comfort in an environment. In this chapter, we will focus only on those specific factors that affect thermal comfort.

Thermal comfort is primarily controlled by a building's heating, ventilating and air-conditioning systems, though the architectural design of the building may also have significant influences on thermal comfort.

This chapter is largely based on ASHRAE's *Standard 55-2004 Thermal Environmental Conditions for Human Occupancy*[1]. In this text, we will abbreviate the title to "*Standard 55*." For a much more in-depth discussion of thermal comfort and the way experimental results are presented, see Chapter 8 of the ASHRAE Handbook, 2005, *Fundamentals*[2].

*Standard 55* defines thermal comfort as "that condition of mind which expresses satisfaction with the thermal environment and is assessed by subjective evaluation." There is no way "state of mind" can be measured. As a result, all comfort data are based on researchers asking questions about particular situations, to build a numerical model of comfort conditions. The model is based on answers to questions by many people under many different experimental conditions.

In the next section, we will consider seven factors influencing comfort and then define acceptable thermal comfort conditions.

## 3.2 Seven Factors Influencing Thermal Comfort

You are a person, so you already know a lot about thermal comfort. You have a lifetime of experience. You know that physical exertion makes you "hot and sweaty." You know you can be more comfortable in a cooler space if you wear more clothes, or warmer clothes. You know that the air temperature matters and that the radiant heat from a fire can help keep you warm and comfortable. You have likely experienced feeling hot in a very humid space and been aware of a cold draft. You have anticipated that a space will be warm and comfortable or cool and comfortable when you get inside.

As a result, you have personal experience of the *seven* factors that affect thermal comfort.

Personal

1. *Activity level*
2. *Clothing*

Individual Characteristics

3. *Expectation*

Environmental Conditions and Architectural Effects

4. *Air temperature*
5. *Radiant temperature*
6. *Humidity*
7. *Air speed*

### 1. Activity Level

The human body continuously produces heat through a process call "metabolism." This heat must be emitted from the body to maintain a fairly constant core temperature, and ideally, a comfortable skin temperature. We produce heat at a minimum rate when asleep. As activity increases, from sitting to walking to running, so the metabolic heat produced increases.

The standard measure of activity level is the "met." One met is the metabolic rate (heat output per unit area of skin) for an individual who is seated and at rest. Typical activity levels and the corresponding met values are shown in *Figure 3.1*.

### 2. Clothing

In occupied spaces, clothing acts as an insulator, slowing the heat loss from the body. As you know from experience, if you are wearing clothing that is an effective insulator, you can withstand, and feel comfortable in lower temperatures.

| Activity | met* |
|---|---|
| Sleeping | 0.7 |
| Reading or writing, seated in office | 1.0 |
| Filing, standing in office | 1.4 |
| Walking about in office | 1.7 |
| Walking 2 mph | 2.0 |
| Housecleaning | 2.0 to 3.4 |
| Dancing, social | 2.4 to 4.4 |
| Heavy machine work | 4.0 |

**Figure 3.1** Typical Metabolic Heat Generation for Various Activities (*Standard 55*, Normative Appendix A, Extracted data) [*1 met = 18.4 Btu/h · ft$^2$]

To predict thermal comfort we must have an idea of the clothing that will be worn by the occupants.

Due to the large variety of materials, weights, and weave of fabrics, clothing estimates are just rough estimates. Each article has an insulating value, unit "clo."

For example: a long-sleeved sweat shirt is 0.34 clo, straight trousers (thin) are 0.15 clo, light underwear is 0.04 clo, ankle-length athletic socks are 0.02 clo, and sandals are 0.02 clo. These clo values can be added to give an overall clothing insulation value. In this case, the preceding set of clothes has an overall clothing insulation value of 0.57 clo.

Typical values for clothing ensembles are shown in *Figure 3.2*. All include shoes, socks, and light underwear.

Later in this Chapter we will introduce a chart, *Figure 3.4*, that illustrates comfortable conditions with 0.5 clo and 1.0 clo. As you can see from *Figure 3.2*, 0.5 clo is very light clothing, and 1.0 clo is heavy indoor clothing.

### 3. Occupants' Expectations

People's expectations affect their perception of comfort in a building. Consider the following three scenarios that all occur on a very hot day:

- A person walks into an air-conditioned office building. The person expects the building to be thermally comfortable.

| Ensemble Description | clo* |
|---|---|
| Trouser, short sleeve shirt | 0.57 |
| Knee-length skirt, short-sleeve shirt (sandals) | 0.54 |
| Trousers, long-sleeved shirt, suit jacket | 0.96 |
| Knee-length skirt, long-sleeved shirt, half slip, panty hose, long-sleeved sweater | 1.10 |
| Long-sleeved coveralls, T-shirt | 0.72 |

**Figure 3.2** Typical Insulation Values for Clothing Ensembles (*Standard 55*, Appendix B, Table B-1, extracted data) [*1 clo = 0.88°F · ft$^2$ · h/Btu]

- A person walks into a prestigious hotel. The person expects it to be cool, regardless of the outside temperature.
- A person walks into an economical apartment building with obvious natural ventilation and open windows. The person has lower expectations for a cool environment. The person anticipates, even hopes, that it will be cooler inside, but not to the same extent as the air-conditioned office building or the hotel.

*Standard 55* recognizes that the expectations for thermal comfort are significantly different in buildings where the occupants control opening windows, as compared to a mechanically cooled building. To address this difference, *Standard 55* provides different criteria for naturally ventilated buildings, as compared to the criteria for mechanically cooled, air-conditioned buildings.

This difference in expectations also shows up in buildings where occupants have a thermostat to control their zone. In general, if occupants have a thermostat in their space, they are more satisfied with their space, even when the performance of the thermostat is very restricted or non-existent (dummy thermostat). This is discussed in the Section 3.3, "Conditions for Comfort."

### *4. Air Temperature*

When we are referring to air temperature in the context of thermal comfort, we are talking about the temperature in the space where the person is located. This temperature can vary from head to toe and can vary with time.

### *5. Radiant Temperature*

**Radiant heat** is heat that is transmitted from a hotter body to a cooler body with no effect on the intervening space. An example of radiant heat transfer occurs when the sun is shining on you. The **radiant temperature** is the temperature at which a black sphere would emit as much radiant heat as it received from its surroundings.

In an occupied space, the floor, walls and ceiling may be at a temperature that is very close to the air temperature. For internal spaces, where the temperature of the walls, floor and ceiling are almost the same as the air temperature, the radiant temperature will be constant in all directions and virtually the same as the air temperature.

When a person is sitting close to a large window on a cold, cloudy, winter day, the average radiant temperature may be significantly lower than the air temperature. Similarly, in spaces with radiant floors or other forms of radiant heating, the average radiant temperature will be above the air temperature during the heating season.

### *6. Humidity*

**Low humidity**: We know that, for some people, low humidity can cause specific problems, like dry skin, dry eyes and static electricity. However, low humidity does not generally cause thermal discomfort. *Standard 55* does not define minimum humidity as an issue of thermal discomfort, nor does it address those individuals who have severe responses to low humidity.

**High humidity**: *Standard 55* does define the maximum humidity ratio for comfort at 0.012 lb/lb. This level of moisture in the air can also cause serious mold problems in the building and to its contents, since it is equivalent to 100% relative humidity at 62°F.

### 7. Air Speed

The higher the air speed over a person's body, the greater the cooling effect. Air velocity that exceeds 40 feet per minute (fpm), or cool temperatures combined with any air movement, may cause discomfort —a draft. Drafts are most noticeable when they blow across the feet and/or the head level, because individuals tend to have less protection from clothing in these areas of their body.

## 3.3 Conditions for Comfort

*Standard 55* deals with indoor thermal comfort in normal living environments and office-type environments. It does not deal with occupancy periods of less than 15 minutes.

The Standard recognizes that individual perceptions of comfort can be significantly modified by prior exposure. For example, consider people coming into a building that is air-conditioned to 82°F on a very hot day, when it is 102°F outside. The building is obviously cooler as they enter it, a pleasant experience. After they have been in the building for half an hour, they will have adjusted and will probably consider the building excessively warm.

When considering issues of comfort, the Standard addresses two situations:

1. **Buildings with occupant-operable windows**
2. **Buildings with mechanically conditioned spaces**

### Situation 1: Buildings with Occupant-Operable Windows

People behave differently when they have windows they can control. They have different, less demanding, expectations due to their knowledge of the external environment and their control over the windows. They will also choose how they dress, knowing that the building temperatures will be significantly influenced by external temperatures.

*Figure 3.3*, shows the **acceptable range** of "indoor operative temperatures" plotted against "mean monthly air temperature" for

   Activity levels of 1.0 to 1.3 met
   Person not in direct sunlight
   Air velocity below 40 fpm
   No specific clothing ensemble values

This acceptable range is called the **comfort envelope**.

The **indoor operative temperature** is the average of the air temperature and radiant temperature.

The **mean monthly outdoor temperature** is the average of the hourly temperatures; data is normally available from government environmental-monitoring departments.

Thermal Comfort 37

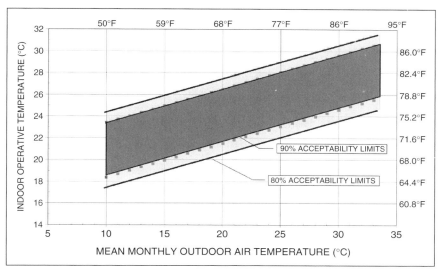

**Figure 3.3** Acceptable Operative Temperature Ranges for Naturally Conditioned Spaces - (*Standard 55*, Figure 5.3)

The chart only goes down to a mean monthly temperature of 50°F, indicating that operant-controlled windows (opening windows) do not provide acceptable thermal comfort conditions in cooler climates during the winter.

The plot shows the range of comfortable operative temperatures for 80% acceptability, the normal situation, and a narrower comfort band that will provide a higher standard of comfort, 90% acceptability. For example, for a location with a maximum summer mean-monthly temperature of 68°F, the range for 80% acceptability is between 71°F and 80°F.

**Note** that the normal situation suggests that 20% of the occupants, or 1 in 5, will not find the thermal conditions acceptable!

### Situation 2: Buildings with Mechanically Conditioned Spaces

Mechanically conditioned spaces are arranged into three classes:

Class A – high comfort
Class B – normal comfort
Class C – relaxed standard of comfort

*Standard 55* includes comfort charts for Class B spaces only. To calculate comfort conditions for Classes A and C, the designer uses a BASIC computer program. The BASIC program listing is included in *Standard 55*, Appendix D.

The Class B thermal limits are based on 80% acceptability, leaving about 10% of the occupants not comfortable due to the overall thermal conditions and 10% not comfortable due to local thermal discomfort.

## Class B Comfort Criteria

The Standard provides a psychrometric chart, *Figure 3.4*, showing acceptable conditions for a Class B space for:

Activity between 1.0 and 1.3 met
Clothing 0.5 to 1.0 clo
The air speed is to be below 40 fpm
The person must not be in direct sunlight

For spaces where it is reasonable to assume that clothing will be around 0.5 clo in the summer, and a design humidity of between 40 and 50%, the acceptable conditions, the comfort envelope, will be within the area shaded with shading lines sloping up to the right.

Remember that the chart is for 80% acceptability, although ideally 100% of the occupants would find the conditions thermally acceptable. The occupants do have some limited flexibility with clothing in most situations. The ideal situation, but prohibitively expensive in most cases, is to provide all the occupants with their own temperature control.

**Example 1**: Let us suppose we wish to minimize the size of the air-conditioning plant; then we could choose design conditions of 81°F at 50% relative humidity, **rh** and 82°F at 40% rh. It must be recognized that when the designer designs on the limit, it means that more people are likely to be uncomfortable than if the designer chooses to design for the center of the comfort temperature band.

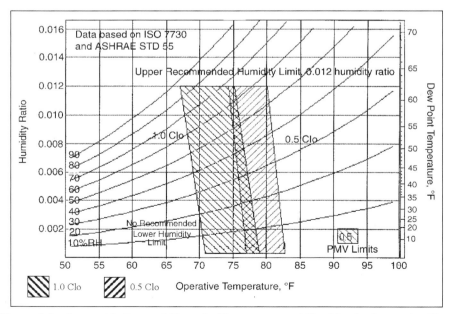

**Figure 3.4** Acceptable Range of Operative Temperature and Humidity for Spaces that Meet the Criteria Specified Above. (*Standard 55*, Figure 5.2.1.1)

**Example 2**: Let us consider a different situation, a prestige office building with, at the design stage, unknown tenants. Here we should allow for both light dress and full suits, the full range 0.5 to 1.0 clo. If the design relative-humidity is to be 50%, then we should select the area of overlap and choose 76°F as our design temperature.

**Example 3**: As a third example let us consider a desert town with an outside design-condition of 90°F and 13% relative humidity. If we pass the incoming air over a suitably sized evaporative cooler, the air will be cooled and humidified to 78°F and 50% which is nicely within the comfort zone for people with 0.5 clo. In this case, we can achieve acceptable thermal comfort for supply ventilation using an evaporative cooler.

## 3.4 Managing Under Less Than Ideal Conditions

The above charts are based on relatively ideal conditions—conditions that do not always exist. The Standard goes into considerable detail about the limits for non-ideal conditions and we will briefly introduce them here.

### Elevated Air Speed

Increasing the air speed over the body causes increased cooling. Elevated air speed can be used to advantage to offset excessive space temperatures. The temperature limits specified are increased by up to 5°F, as long as the air speed is within the occupant's control and limited to below 160 ft/min.

The personal desk fan provides a simple example of placing air speed under individual control. For example, in the case of a naturally ventilated space where the acceptable temperature range was 71°F to 80°F, the acceptable temperature range would be increased to a higher range of 71°F to 85°F with the addition of a fan that was controlled by the occupant.

### Draft

Draft discomfort depends on air temperature, velocity and turbulence. In general the steadier the draft the less the discomfort—it does not draw attention to itself so much! People are much more sensitive to cold drafts than they are to warm drafts. As a result the same velocity of air may produce complaints of cold drafts while cooling in the summer but no complaints when heating in the winter.

### Vertical Temperature Difference

Vertical temperature difference between feet and head typically occurs in heated buildings. Warm air is less dense and tends to rise. Therefore, a warm air supply tends to rise, leaving the lower portion of the space cooler.

Also, many buildings in cool climates have a poorly insulated floor slab-on-grade, which makes for a cold floor and cool air just above the floor.

The variation in air temperature from feet to head is generally acceptable as long as it does not exceed 5°F.

*Floor Surface Temperatures*

Floor surface temperatures should be within the range 66–84°F for people wearing shoes and not sitting on the floor. The maximum temperature limits the amount of heat that can be provided by a heated (radiant) floor. The minimum temperature, 66°F, is much higher than most designers realize! Note that a cold floor can make it impossible to produce thermal comfort, regardless of the temperature of the space.

*Cyclic Temperature Changes*

In a space that is controlled by an on/off thermostat that reacts slowly to temperature change, the space can experience a significant temperature range in a short time. The occupants can perceive this variation as discomfort.

When the temperature cycles up and down fairly regularly with time, with a cycle time of less than 15 minutes, the temperature range should be limited to a range of 2°F.

*Radiant Temperature Variation*

Radiant temperature variation is acceptable, within limits. People are generally quite accepting of a warm wall, but warm ceilings are a source of discomfort if the ceiling radiant temperature is more than 9°F above the general radiant temperature.

A poorly insulated roof in a hot sunny climate can cause very uncomfortable conditions due to the high radiant temperature of the ceiling.

## 3.5 Requirements of Non-Standard Groups

This has been a very brief look at the variations in thermal conditions that can influence the basic comfort charts in *Figures 3.1* and *3.2*. There has been no mention of different requirements for different age groups or sexes. Most research is done on healthy adults, and *Standard 55* admits this fact by noting that there is little data on the comfort requirements for children, the disabled or the infirm.

However, most research on differences between groups indicates that different acceptability is due to different behavior, rather than different thermal comfort requirements. For example, elderly people often like a warmer temperature than younger people do. This is reasonable, since the elderly tend to be much less active, resulting in a lower met rate. In a similar way, women are thought to prefer a warmer temperature than men, but comparative studies indicate that the reason for the difference is that women wear a lower clo value ensemble of clothes.

Lastly there is the idea that people prefer their space to be cooler in summer and warmer in winter. Consider a one-level house. In summer, it is hot and sunny outside. As a result, the walls and roof become much warmer than they are in cooler weather. For the occupant, the radiant temperature is higher, and therefore, to maintain the same thermal conditions, the air temperature needs to be lower. Conversely, in cold winter weather, the walls, windows and ceilings become cooler and the occupant will need a higher air temperature to maintain the same level of comfort.

# The Next Step

Having considered thermal comfort in this chapter we will go on to consider indoor air comfort, termed *Ventilation and Indoor Air Quality*, in Chapter 4.

# Summary

This chapter has considered the many facets of thermal comfort. It is important that you are aware that the air temperature at the thermostat is not always a good indicator of thermal comfort. The design of the space and individual clothing choices can have major influences on thermal comfort.

### Section 3.1 Introduction – What is Thermal Comfort?

*Standard 55* defines comfort as "that condition of mind which expresses satisfaction with the thermal environment; it requires subjective evaluation."

### Section 3.2 Seven Factors influencing Comfort

You have personal experience of the seven factors that affect thermal comfort: personal comfort, including activity level and clothing; individual characteristics, including expectation; environmental conditions and architectural effects, including air temperature, radiant temperature, humidity, and air speed.

### Section 3.3 Conditions for Comfort

This section focuses on the factors that influence thermal comfort in normal living environments and office-type environments with occupancy periods in excess of 15 minutes. These include occupant operable windows and naturally conditioned spaces, and mechanically conditioned spaces. Mechanically conditioned spaces are arranged into three classes: Class A – high comfort; Class B – normal comfort; Class C – relaxed standard of comfort. The Standard provides a psychrometric chart showing 80% acceptable conditions for a Class B space for activity between 1.0 and 1.3 met; clothing 0.5 to 1.0 clo; air speed below 40 fpm; with the added condition that the person is not in direct sunlight. To calculate comfort conditions for Classes A and C, the designer uses a BASIC computer program.

### Section 3.4 Managing Under Less Than Ideal Conditions

Non-ideal conditions include: elevated air speed, draft, vertical temperature difference, floor surface temperatures, cyclic temperature changes, and radiant temperature variation.

### Section 3.5 Requirements of Non-Standard Groups

Most of the research for *Standard 55* was based on the responses of healthy adults. When designing for non-standard groups, consider their additional needs for comfort.

## Bibliography

1. ASHRAE, *Standard 55-2004 Thermal Environmental Conditions for Human Occupancy*
2. ASHRAE Handbook, *Fundamentals*, 2005

Chapter 4

# Ventilation and Indoor Air Quality

## Contents of Chapter 4

Study Objectives of Chapter 4
4.1 Introduction
4.2 Air Pollutants and Contaminants
4.3 Indoor Air Quality Effects on Health and Comfort
4.4 Controling of Indoor Air Quality
4.5 ASHRAE Standard 62 Ventilation for Acceptable Indoor Air Quality
The Next Step
Summary
Bibliography

## Study Objectives of Chapter 4

Chapter 4 deals with the reasons for ventilating buildings and how ventilation rates are chosen for specific situations. After studying the chapter, you should be able to:

List, and give examples of the four types of indoor air contaminants
Describe the three methods of maintaining indoor air quality
Understand the criteria for filter selection
Understand the main concepts of the ASHRAE Standard 62.1-2004 ventilation rate procedure and how it differs from ASHRAE Standard 62.1-2001

## 4.1 Introduction

In Chapter 3, we covered two factors that affect comfort and activity, temperature and humidity. In this chapter, we will be discussing an additional factor, Indoor Air Quality, IAQ. The maintenance of indoor air quality (IAQ) is one of the major objectives of air-conditioning systems because IAQ problems are a significant threat to health and productivity.

Those who study Indoor Air Quality consider the makeup of indoor air, and how it affects the health, activities and comfort of those who occupy the space.

The primary factors that influence and degrade IAQ are particles, gases, and vapors in the air. Maintenance of good indoor air quality is a significant issue to both the HVAC design engineer and to those who maintain the system subsequent to its design and installation.

To deal properly with the issues of IAQ, it is important to be aware of

**The various types of pollutants and contaminants, their sources and their effects on health.**
**The factors that influence pollutant and contaminant levels in buildings**
- The sources of pollutants.
- The ways pollutants can be absorbed and re-emitted into the building spaces.

**Ways of maintaining good IAQ by**
- Controlling the source of pollutants within the space.
- Using filters to prevent pollutants and contaminants from entering the space.
- Diluting the pollutants and contaminants within the space.

ASHRAE has two ANSI approved standards on ventilation:

ANSI/ASHRAE Standard 62.1-2004, *Ventilation for Acceptable Indoor Air Quality*[1] (Standard 62.1-2004) which deals with ventilation in "all indoor or enclosed spaces that people may occupy."
ASHRAE/ANSI Standard 62.2-2004 *Ventilation and Acceptable Indoor Air Quality in Low Rise Residential Buildings*[2] (Standard 62.2) which deals, in detail, with residential ventilation.

The scope of Standard 62.1-2004 deals specifically with "Release of moisture in residential kitchens and bathrooms," while Standard 62.2 deals with "mechanical and natural ventilation systems and the building envelope intended to provide acceptable indoor air quality in low-rise residential buildings."

Like other ASHRAE standards, these are consensus documents, produced by a volunteer committee of people who are knowledgeable in the field. The standards have been publicly reviewed and are continuously re-assessed. They have force of law only when adopted by a regulatory agency, but are generally recognized as being the standard of minimum practice.

## 4.2 Air Pollutants and Contaminants

Air pollutants and contaminants are unwanted airborne constituents that may reduce the acceptability of air. The number and variety of contaminants in the air is enormous. Some contaminants are brought into the conditioned space from outside, and some are generated within the space itself. *Figure 4.1* lists some of the most common indoor air contaminants and their most common sources.

| Contaminants | Major Source |
|---|---|
| *Particles (particulates)* | Dust (generated inside and outside), smoking, cooking |
| Allergens (a substance that can cause an allergic reaction) | Molds, pets, many other sources |
| Bacteria and Viruses | People, moisture, pets |
| Carbon Dioxide ($CO_2$) | Occupants breathing, combustion |
| Odoriferous chemicals | People, cooking, molds, chemicals, smoking |
| Volatile Organic Compounds (VOCs) | Construction materials, furnishings, cleaning products |
| Tobacco Smoke | Smoking |
| Carbon Monoxide (CO) | Incomplete and/or faulty combustion, smoking |
| Radon (Rn) | Radioactive decay of radium in the soil |
| Formaldehyde (HCHO) | Construction materials, furniture, smoking |
| Oxides of Nitrogen | Combustion, smoking |
| **Sulphur Dioxide** | Combustion |
| Ozone | Photocopiers, electrostatic air cleaners |

**Figure 4.1** Common Air Contaminants

## 4.3 Indoor Air Quality Effects on Health and Comfort

It is important to distinguish between the various contaminants in terms of their health effects. The HVAC designer and building operator may take different approaches to contaminants that can be detrimental to health and those that are merely annoying. Although it is the annoying aspects that will draw immediate attention from the occupants, it is the health affecting contaminants that are of the utmost short and long term importance. It is useful to think of contaminants in terms of the following classes of effect:

*Fatal in the short term*
*Carcinogenic (cancer causing substances)*
*Health threatening*
*Annoying, with an impact on productivity and sense of well-being*

### Fatal in the Short Term

At times, contaminants are found in buildings in concentrations that can cause death. These include airborne chemical substances, such as carbon monoxide, or disease-causing bacteria and other biological contaminants.

Carbon monoxide, a colorless and odorless gas, is produced during incomplete combustion. It is attributed as the cause of many deaths each year. One source of carbon monoxide is a malfunctioning combustion appliance, such as a furnace, water heater or stove. Another possible source of carbon monoxide is the exhaust that results from operating a combustion engine or motor vehicle in an enclosed space.

Certain disease-causing bacteria can be present in the air in the building. These include contagious diseases, such as tuberculosis, exhaled by people who are infected with the disease. The tubercle bacillus is very small and tend to stay afloat in the air. Exposure can be minimized by isolating affected individuals, and by using special ventilation methods.

A third group of contaminants are disease causing bacteria that are generated by physical activity or equipment. One, which is particularly dangerous for people with a weak immune system, is legionella. Legionella is the bacteria that causes Legionnaire's Disease. Legionella multiplies very rapidly in warm, impure water. If this water is then splashed or sprayed into the air, the legionella bacteria become airborne and can be inhaled into a person's lungs. Once in the lungs, the bacteria pass through the lung wall and into the body. The resultant flu-like disease is often fatal.

The source of a legionella outbreak can often be traced to a particular location, such as a cooling tower or a domestic hot water system. Where we know the source and the mechanism of transfer of disease to the individual, we call it a "building related illness."

The pollutants that are fatal in the short-term are often unnoticeable except as a result of their health effects.

### *Carcinogens*

Carcinogens are among the most significant contaminants because of their potential to cause cancer in the long-term. The risk of cancer increases with level and time of exposure to the substance. The exposure may be unnoticeable and not have any immediately apparent impact in the short-term. However, in the long-term, even low levels of exposure may lead to severe, irreversible health problems.

Environmental tobacco smoke (ETS) has been one of the major concerns in maintaining good indoor air quality. Concern has been heightened by increased evidence of its role in lung and heart disease. Most tobacco-related deaths occur among the smokers themselves, but tobacco smoke in the indoor air can also cause cancer in non-smokers. The smoke also causes physical irritation, annoyance and dirt on all exposed surfaces.

Another carcinogen of concern in some places is the gas radon. Radon is a naturally occurring radioactive gas that results from the decay of radium in the soil. This radioactive gas leaks into buildings where it can be inhaled and potentially cause cancer. In places where radon is an issue, it can be controlled by venting the crawlspace, sealing all cracks, or by pressurizing the interior so as to minimize radon entry.

### *Health Threatening*

Many indoor air contaminants (such as allergens, volatile organic compounds, bacteria, viruses, mold spores, ozone and particulates) can be physically irritating or health threatening, although they are not usually fatal. Among the most common symptoms is the irritation of delicate tissues such as the eyes, skin, or mucous membranes. Many contaminants cause cold-like symptoms that are often mistaken as the effects of a viral infection.

In some buildings, a significant proportion of the occupants may experience symptoms. If the symptoms disappear when the occupants have left the building, one can surmise that something in the building is causing the symptoms.

If 20% or more of the occupants experience the symptoms *only* when they are in the building, then they are considered to be suffering from "sick building syndrome."

### Annoying, with an Impact on Productivity and Sense of Well-Being

Although not health threatening, many odoriferous chemicals are annoying and may be distracting enough to affect productivity without threatening health. These include body odors, some chemicals, the smells of spoiling food, and some molds that do not have more serious effects. In high enough concentrations, some contaminants have physical effects that are gradual and subtle enough not to be immediately noticed.

## 4.4 Controlling Indoor Air Quality

Maintaining acceptable IAQ depends on the judicious use of three methods:

Source control
Filtration
Dilution

### 4.4.1 Source Control

The most important method of maintaining acceptable indoor air quality is by controlling sources of contaminants and pollutants. Sources can be controlled by restricting their access to the space, either by design or by appropriate maintenance procedures, and by exhausting pollutants that are generated within the space. Avoiding the use of volatile solvents and banning smoking are two simple indoor examples.

Another example of source control is found in a new requirement in *Standard 62.1-2004* where it states that water for humidifiers "shall originate directly from a potable source or from a source with equal or better water quality." In the past, steam from the steam heating system was often used for humidification of buildings. This steam was frequently treated with anticorrosion additives that would not be acceptable in potable water. Now, this steam is not an acceptable source for direct humidification.

When designing the air intake system, one should always deliberately reduce the likelihood of pollutants coming in from outside. Methods include locating intakes:

Away from the ground, where dust blows by
Away from loading docks, where there are higher concentrations of pollutants from vehicles
Away from outlets on the roof that vent things, such as toilets, furnaces, drains, and fume hoods

One common source of indoor pollution is mold. The spores and dead particles of mold adversely affect many people. To prevent mold, keep the building fabric and contents reasonably dry. As a general rule, maintain the relative humidity below 60% to prevent mold growth. This is a challenge in a hot, humid climate with air-conditioned buildings where the outside air contains so much moisture. For example, a new prestigious hotel in Hawaii had to be closed within a year of opening, due to mold in over 400 bedrooms. Remedial costs will exceed $US10 million.

One source of mold, that is often neglected, is the drain pan beneath a cooling coil. The coil collects moisture and, being wet, some dirt out of the air. Ideally, this moisture and dirt drips down into the tray and drains away. Unfortunately, (and frequently), if the tray has a slope-to-drain ratio that is less than the required 1/8 inch per foot, a layer of sludge can form in the tray and grow mold. If the coil is not used for cooling for a while, the tray dries out and the crust of dried sludge can breakup and get carried through the system into the occupied spaces. Regular cleaning of the tray is required to minimize the problem.

If the pollution is from a specific source indoors, then direct exhaust can be used to control the pollutants. For example: the hood over a cooking range pulls fumes directly from the stove and exhausts them; exhausting the fumes from large photocopiers avoids contaminating the surrounding office space; and the laboratory fume cabinet draws chemical fumes directly to outside. When designing any direct exhaust system, one should attempt to collect the pollutant before it mixes with much room air. This reduces the required exhaust air volume and hence reduces the amount of conditioned air required to make up for the exhaust.

The design of exhaust systems for a large variety of situations is very clearly explained and accompanied with explanatory diagrams in *Industrial Ventilation*[3], published by the American Conference of Governmental Industrial Hygienists.

**4.4.2 Filtration**

Filtration is the removal of contaminants from the air. Both particulate (particles of all sizes) and gaseous contaminants can be removed, but since gaseous filtration is a rather specialized subject, we will not discuss it in this course.

Particulate filters work by having the particles trapped by, or adhere to, the filter medium. The actual performance of a filter depends on several factors, including particle size, air velocity through the filter medium, filter material and density, and dirt buildup on the filter. The main operating characteristics used to distinguish between filters are:

Efficiency in removing dust particles of varying sizes
Resistance to airflow
Dust-holding capacity (weight per filter)

Choosing a filter is a matter of balancing requirements against initial purchase cost, operating cost and effectiveness. In general, both the initial cost and the operating cost of the filter will be affected by the size of the particles that need to be filtered out, and the required efficiency of the filter: the smaller the particle size and the greater the efficiency required, the more expensive the filter costs.

The *Figure 4.2* shows a sample of particles and their range of size.

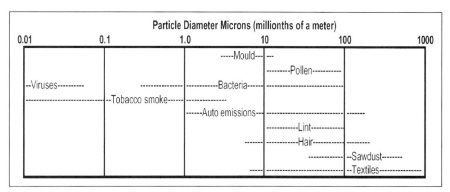

**Figure 4.2** Particle Diameter, Microns (millionths of a meter)

Information on filter performance is usually based on a standard. For the HVAC industry, ASHRAE has produced two standards. The first was ASHRAE Standard 52.1-1992 *Gravimetric and Dust Spot Procedures for Testing Air Cleaning Devices used in General Ventilation for Removing Particulate Matter*[4] (Standard 52.1). Testing a filter to Standard 52.1 produces an "ASHRAE atmospheric dust spot efficiency" and an "ASHRAE arrestance." The "dust spot" efficiency is a measure of how well the filter removes the finer particles that cause staining, and the "arrestance" is a measure of the weight of dust that is collected before the resistance of the filter rises excessively. Unfortunately, the dust spot efficiency does not give much information about filter performance for different particle sizes and does not differentiate among less efficient filters.

As a result, a new standard was introduced, ASHRAE 52.2-1999 *Method for Testing General Ventilation Air-Cleaning Devices for the Removal Efficiency by Particle Size*[5]. It is based on using a particle counter to count the number of particles in twelve different size fractions. This data is used to classify a filter into one of 20 "Minimum Efficiency Reporting Values" called MERV. The least efficient filter is MERV 1 and the most efficient, MERV 20. *Figure 4.3* shows typical filters with their range of performance and typical applications.

There are numerous types of filters, made with a variety of filter media. The simplest, cheapest, and generally least effective, is the panel filter. The panel filter, commonly used in residential systems, is a pad of filter media across the air stream. The pad can be aluminum mesh, to provide a robust washable unit, typically having a MERV rating 1 to 3. The media may be a bonded fiberglass cloth with a MERV rating up to 4. There are many other constructions that are designed to satisfy the market at an affordable price.

The performance of the panel filter can be improved by mounting panel filters at an angle to the air stream to form an extended surface. For the same air velocity through the duct, the filter area is increased and the velocity through the media is decreased to improve performance.

The filtering performance and dust holding capacity can be further improved by pleating the media. Variations of pleated media filters cover the MERV range from 5 to 8.

To achieve a higher dust holding capacity, the media can be reinforced and formed into bags of up to 36 inches deep. The bags are kept inflated by the flow of air through them during system operation.

These arrangements are shown diagrammatically in *Figure 4.3*.

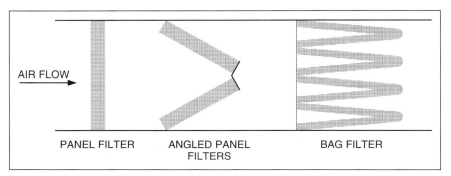

**Figure 4.3** Basic Filter Media Filter Arrangements

Two of the factors that influence filter performance are the filter media and the air velocity through the media. Some filters have graded media with a coarse first layer to collect most of the large particles, and then one or more finer layers to catch progressively smaller particles. As a result of the grading, the final fine layer does not get quickly clogged with large particles. Pleated and bag filters extend the surface of the filter. This reduces the velocity of the air through the fabric and greatly increases the collection area for the particles, resulting in a much higher dust-holding capacity.

For ventilation systems, filters with a MERV above 8 are almost always provided with a pre-filter of MERV 4 or less to catch the large particles, lint and insects. It is more economical to remove the large particles with a course filter and prolong the life of the better filter.

Electronic filters can be used as an alternative to the media filters discussed above. In an electronic filter, the air passes through an array of wires. The wires are maintained at a high voltage, which generates an electrical charge on the dust particles. The air then passes on between a set of flat plates that alternate between high voltage and low voltage. The charged dust particles are attracted to the plates and adhere to them. These filters can be very efficient but they require cleaning very frequently to maintain their performance. Larger systems often include automatic wash systems to maintain the performance.

### Filter Characteristics

Let us return to the three main filter characteristics:

Efficiency in removing dust particles of varying sizes
Resistance to airflow
Dust-holding capacity

*Efficiency in removing dust particles of varying sizes* is influenced by how clean the space is required to be, and whether any particular particles are an issue. Thus one might choose a MERV 5 to 8 filter in an ordinary office building, but a MERV 11 to 13 filter in a prestige office complex. The higher MERV filters cost more to install and to operate but they reduce dirt in the building and so they save on cleaning and redecorating costs.

## Ventilation and Indoor Air Quality

| Standard 52.2 Minimum Efficiency Reporting Value (MERV) | Approximate Standard 52.1 Results | | Application Guidelines | | |
|---|---|---|---|---|---|
| | Dust Spot Efficiency | Arrestance | Typical Controlled Contaminant | Typical Applications and Limitations | Typical Air Cleaner/ Filter Type |
| 20<br>19<br>18<br>17 | n/a<br>n/a<br>n/a<br>n/a | n/a<br>n/a<br>n/a<br>n/a | **Larger than 0.3 $\mu$m particles**<br>Virus<br>All combustion smoke<br>Sea salt<br>Radon progeny | Cleanrooms<br>Pharmaceutical manufacturing<br>Orthopedic surgery | HEPA/ULPA filters ranging from 99.97% efficiency on 0.3 $\mu$m particles to 99.999% efficiency on 0.1–0.2 $\mu$m particles |
| 16<br>15<br>14<br>13 | n\a<br>>95%<br>90–95%<br>80–90% | n/a<br>n/a<br>>98%<br>>98% | **0.3–1.0 $\mu$m Particle size, and all over 1 $\mu$m**<br>All bacteria<br>Most tobacco smoke<br>Sneeze nuclei | Hospital inpatient care<br>General surgery<br>Superior commercial buildings | **Bag filters**<br>Nonsupported (flexible) microfine fiberglass or synthetic media 12 to 36 inches deep, 6 to 12 pockets |
| 12<br>11<br>10<br>9 | 70–75%<br>60–65%<br>50–55%<br>40–45% | >95%<br>>95%<br>>95%<br>>90% | **1.0–3.0 $\mu$m Particle size, and all over 3.0 $\mu$m**<br>Legionella<br>Auto emissions<br>Welding fumes | Hospital laboratories<br>Better commercial buildings<br>Superior residential | **Box filters**<br>Rigid style cartridge filters 6 to 12 inches deep may use lofted (air laid) or paper (wet laid) media |
| 8<br>7<br>6<br>5 | 30–35%<br>25–30%<br><20%<br><20% | >90%<br>>90%<br>85–90%<br>80–85% | **3.0–10.0 $\mu$m Particle size, and all over 10 $\mu$m**<br>Mold<br>Spores<br>Cement dust | Commercial buildings<br>Better residential<br>Industrial workplaces | **Pleated filters**<br>Disposable extended surface, 1 to 5 inch thick with cotton-polyester blend media, cardboard frame<br>**Cartridge filters**<br>Graded density viscous coated cube or pocket filters, synthetic media<br>**Throwaway** Disposable synthetic media panel filters |
| 4<br>3<br>2<br>1 | <20%<br><20%<br><20%<br><20% | 75–80%<br>70–75%<br>65–70%<br><65% | **>10.0 $\mu$m Particle size**<br>Pollen<br>Dust mites<br>Sanding dust<br>Textile fibers | Minimum filtration<br>Residential<br>Window air conditioners | **Throwaway** Disposable fiberglass or synthetic panel filters<br>**Washable** Aluminum mesh, latex coated animal hair, or foam rubber panels<br>**Electrostatic** Self charging (passive) woven polycarbonate panel filter |

**Figure 4.4** Filter Test Performance and Applications (extracted from ASHRAE Standard 52.2-1999, Page 39)

When it comes to medical facilities, MERV 14 to 16 filters will remove most bacteria and can be used for most patient spaces. For removal of all bacteria and viruses, a MERV 17, called a **HEPA filter**, is the standard filter. HEPA filters have an efficiency of 99.7% against 0.3 micron particles.

*Resistance to airflow* directly affects the fan horsepower required to drive the air through the filter. Many less expensive, pre-packaged systems do not have fans that are capable of developing the pressure to drive air through the dense filter material of the higher MERV rated filters. Typically, most domestic systems will handle the pressure drop of a MERV 5 or 6 filter, but not higher.

*Dust-holding capacity* influences the filter life between replacements. A pleated filter with MERV 7 or 8 may be all that is required, but a bag filter with MERV 9 or 10 can have a much higher dust holding capacity. The bag filter could, therefore, be a better choice in a very dirty environment or where there is a high cost to shut down the system and change the filters.

### 4.4.3 Dilution

In most places the outside air is relatively free of pollutants, other than large dust particles, birds, and insects. When this air is brought into a space, through a screen and filter to remove the coarse contaminants, it can be used to dilute any contaminants in the space. We also need a small supply of outside air to provide us with oxygen to breathe and to dilute the carbon dioxide we exhale. Dilution ventilation is the standard method of controlling general pollutants in buildings and the methods and quantities required are detailed in Standard 62.1-2004, which is the subject of the next section, 4.5.

## 4.5 ASHRAE Standard 62 Ventilation for Acceptable Indoor Air Quality

ANSI/ASHRAE Standard 62, *Ventilation for Acceptable Indoor Air Quality*[1] was published in 1971, 1981 and again fully revised in 1989. The complete revisions made it easy to reference in Building Codes. Designers could refer to the edition stipulated, and there was no question about the reference. The policy was changed for this standard in 1997, to align with the ANSI "continuous maintenance" process. Under continuous maintenance, the Standard is updated a bit at a time and is not required to be a consistent, whole document. The information in this section is based on the 2004 printed edition.

Standard 62.1-2004 applies to "all indoor or enclosed spaces that people may occupy" with the provision that additional requirements may be necessary for laboratory, industrial, and other spaces. As noted at the beginning of this chapter in the introduction, residential ventilation is specifically covered in *Standard 62.2-2004 Ventilation and Acceptable Indoor Air Quality in Low-Rise Residential Buildings*. You should also note that many local authorities have more demanding and specific requirements for residential ventilation than the ASHRAE standards. For industrial occupancies, refer to *Industrial Ventilation*, published by the American Conference of Governmental Industrial Hygienists.

The first section of Standard 62.1-2004 states

> "The purpose of this standard is to specify minimum ventilation rates and indoor air quality that will be acceptable to human occupants and are intended to minimize the potential for adverse health effects."

Note that this is a minimum standard, that it is aimed at providing "acceptable indoor air quality" which is defined as:

> "air in which there are no known contaminants at harmful concentrations as determined by cognizant authorities and with which a substantial majority (80% or more) of the people exposed do not express dissatisfaction."

The Standard defines two types of requirements to maintain indoor air quality: requirements to limit contamination; and requirements to provide ventilation to dilute and remove contaminants. The requirements to limit contamination also include several system and building design requirements to minimize moisture problems that typically lead to mold problems including:

Requirements for filtering
Separation distance between outside air inlets and contaminated exhausts
Rules about recirculation of air between zones that have different contamination levels
Requirements for maintenance and operation
Requirements for design and maintenance documentation

Standard 62.1-2004 requires that "Air from smoking areas shall not be recirculated or transferred to no-smoking areas." Also smoking areas "shall have more ventilation and/or air cleaning than comparable no-smoking areas." However no specific recommendations are included for smoking areas.

There are two approaches to providing ventilation for the occupants to breathe and to dilute the inevitable pollutants:

- "The *Indoor Air Quality Procedure*" Acceptable air quality is achieved within the space by controlling known and specifiable contaminants to acceptable limits. The application of the Indoor Air Quality Procedure allows the use of particulate and gaseous filters to assist in maintaining acceptable indoor air quality. The complexity of the procedure is beyond this course and will not be discussed.
- "The *Ventilation Rate Procedure*" Acceptable air quality is achieved by providing ventilation air of the specified quality and quantity.

The Ventilation Rate Procedure is based on providing an adequate supply of acceptable outdoor air to dilute and remove contaminants in the space to provide acceptable IAQ. Acceptable outdoor air must have pollution levels within national standards.

The basic required outside air for ventilation is based on a rate, cfm, per person, plus a rate per square foot, $cfm/ft^2$. This basic requirement is then adjusted to allow for the ventilation effectiveness in each space and the

### TABLE 6-1 MINIMUM VENTILATION RATES IN BREATHING ZONE
(This table is not valid in isolation; it must be used in conjunction with the accompanying notes.)

| Occupancy Category | People Outdoor Air Rate $R_p$ | | Area Outdoor Air Rate $R_a$ | | Notes | Default Values | | | | Air Class |
|---|---|---|---|---|---|---|---|---|---|---|
| | | | | | | Occupant Density (see Note 4) | | Combined Outdoor Air Rate (see Note 5) | | |
| | cfm/ person | L/s· person | cfm/ft² | L/s·m² | | #/1000 ft² or #/100 m² | | cfm/ person | L/s· person | |
| **Hotels, Motels, Resorts, Dormitories** | | | | | | | | | | |
| Bedroom/living Room | 5 | 2.5 | 0.06 | 0.3 | | 10 | | 11 | 5.5 | 1 |
| Barracks sleeping areas | 5 | 2.5 | 0.06 | 0.3 | | 20 | | 8 | 4.0 | 1 |
| Lobbies/prefunction | 7.5 | 3.8 | 0.06 | 0.3 | | 30 | | 10 | 4.8 | 1 |
| Multi-purpose assembly | 5 | 2.5 | 0.06 | 0.3 | | 120 | | 6 | 2.8 | 1 |
| **Office Buildings** | | | | | | | | | | |
| Office space | 5 | 2.5 | 0.06 | 0.3 | | 5 | | 17 | 8.5 | 1 |
| Reception areas | 5 | 2.5 | 0.06 | 0.3 | | 30 | | 7 | 3.5 | 1 |
| Telephone/data entry | 5 | 2.5 | 0.06 | 0.3 | | 60 | | 6 | 3.0 | 1 |
| Main entry lobbies | 5 | 2.5 | 0.06 | 0.3 | | 10 | | 11 | 5.5 | 1 |

GENERAL NOTES FOR TABLE 6–1

1 **Related Requirements:** The rates in this table are based on all other applicable requirements of this standard being met.
2 **Smoking:** This table applies to no-smoking areas. Rates for smoking-permitted spaces must be determined using other methods. See Section 6.2.9 for ventilation requirements in smoking areas.
3 **Air Density:** Volumetric airflow rates are based on an air density of 0.075 $lb_{da}/ft^3$ (1.2 $kg_{da}/m^3$), which corresponds to dry air at a barometric pressure of 1 atm (101.3 kPa) and an air temperature of 70°F (21°C). Rates may be adjusted for actual density but such adjustment is not required for compliance with this standard.
4 **Default Occupant Density:** The default occupant density shall be used when actual occupant density is not known.
5 **Default Combined Outdoor Air Rate (per person):** This rate is based on the default occupant density.
6 **Unlisted Occupancies:** If the occupancy category for a proposed space or zone is not listed, the requirements for the listed occupancy category that is most similar in terms of occupant density, activities and building construction shall be used.
7 **Residential facilities, Healthcare facilities and Vehicles:** Rates shall be determined in accordance with Appendix E.

**Figure 4.5**  Parts of Table 6-1, ASHRAE Standard 62.1-2004

effectiveness of the system. Let us briefly go through those steps. An excerpt of the base ventilation data from Table 6–1 in Standard 62.1-2004 is shown in *Figure 4.5*.

Look at the first occupancy category, the hotel bedroom. The requirement is here is for 5 cfm per person and 0.06 cfm/ft². Based on the default occupancy density of 10 persons per 1000 ft² the combined outdoor rate per 1000 ft² is

$$10 \text{ people} \cdot 5 \text{ cfm/person} + 1000 \text{ ft}^2 \cdot 0.06 \text{ cfm/ft}^2 = 50 \text{ cfm} + 60 \text{ cfm} = 110 \text{ cfm}$$

The default combined outdoor air rate is thus 110 cfm for 10 people occupying 1000 ft2. Divided by the default population of 10 persons we get 11 cfm/person for the base requirement per person.

Now look at the last hotel category, multi-purpose assembly. The rate per person, 5 cfm, and rate per ft$^2$, 0.06 cfm, are the same. What is different is the default occupancy density of 120 persons/1000 ft$^2$. With the much higher occupancy density the ventilation for the space is much less significant and therefore the combined outdoor air rate per person is halved to 5.5 cfm, shown rounded up to 6 cfm in the table.

These default outdoor air rates must then be adjusted to allow for the proportion of ventilation air that actually circulates through the breathing zone. If we suppose that only 90% of the outdoor air enters the breathing zone, and the other 10% circulates above the breathing zone and is exhausted, then only the 90% of outside air is being used effectively. Therefore, the proportion of air that actually circulates into the breathing zone is called **zone air distribution effectiveness**. In the example, the zone air distribution effectiveness would be 0.9. The **breathing zone** is defined as between 3 and 72 inches from the floor and 24 inches from walls or air-conditioning equipment.

Let us consider a space with the ventilation air being provided from a ceiling outlet. Standard 62.1-2004 gives the zone air distribution effectiveness for cool air supplied at ceiling level as "1." To obtain the corrected ventilation rate, we divide the base rate by the zone air distribution effectiveness. In this case, default outdoor air rate divided by a zone air distribution effectiveness of "1" means the default rate is unchanged.

Now let us suppose that the same system is used for heating in the winter. In this example, the maximum design supply temperature is 95°F and space design temperature is 72°F. The supply air temperature is

$95°F - 72°F = 23°F$

above the temperature of the space. According to Standard 62.1-2004, "For warm air over 15°F above space temperature supplied at ceiling level and ceiling return, the zone air distribution effectiveness is 0.8." In this example, with the default rate divided by 0.8, we obtain the corrected required ventilation, $1/0.8 = 1.25$. This means that the outside air requirement has increased by 25%, compared to the cooling-only situation. If this system runs all year, then the ventilation should be designed for the higher winter requirement.

Thus far, we have used the Table 6-2 rates to obtain base ventilation rates and then corrected those to recognize zone air distribution effectiveness within the space. Now we must consider the effectiveness of the system.

If the system supplies just one zone or 100% outside air to several zones, the calculated rate is used. However, if the system serves multiple zones with a mixture of outside air and recirculated return air, we may have to make a system adjustment to allow for differing proportions of outside air going to different zones.

For example, an office building might require 15% outside air to the offices, but 25% to the one conference room. If the system provides only 15%, then the conference room will be under-ventilated. However, 25% for the conference room will provide much more than the required ventilation

56   Fundamentals of HVAC

to the rest of the offices. Standard 62.1-2004 includes a simple calculation to obtain a rate between 15% and 25% that provides adequate outside air for all the zones.

Further adjustments can be made to allow for variable occupancy and for short interruptions in system operation. Just one example of this type of adjustment can occur in churches with high ceilings. If the services are of limited duration, say under an hour and a half, and the volume of the zone is large per person, then the outside air ventilation rate can be based on an average population over a calculated period. This may substantially reduce the required flow of outside air.

This discussion has all been based on Standard 62.1-2004. In many jurisdictions, earlier versions of the standard will remain the legal requirement for many years. If this is the case in your jurisdiction, it is important to know that previous versions of the Standard generally calculated the required ventilation based on cfm-per-person and took no separate account of the size of the zone. The simpler requirement facilitated a simple method of adjusting ventilation rates to meet actual occupancy needs in densely occupied spaces. The following section describes how carbon dioxide can be used to determine ventilation requirements in these situations.

### 4.5.1 The Use of Carbon Dioxide to Control Ventilation Rate

All versions of the Standard allow for reduced ventilation when the population density is known to be lower. For example, the ventilation for a movie theatre must be sized for full occupancy, although the theatre may often be less than half-full. In these "less-than-full" times it would save energy if we could reduce the ventilation rate to match the actual population. In the versions of Standard 62 that preceded 2004, the ventilation rates were based on cfm/person. As a result, the ventilation could be adjusted based on the number of people present.

Conveniently for the purposes of measurement, people inhale air that contains oxygen and exhale a little less oxygen and some carbon dioxide. The amount of carbon dioxide, $CO_2$, that is exhaled is proportional to a person's activity: more $CO_2$ is exhaled the more active the person. This exhaled $CO_2$ can be measured and used to assess the number of people present.

In our movie theatre, the people (assume adults) are all seated and the metabolic rate is about 1.2 met. At 1.2 met, the average $CO_2$ exhaled by adults is 0.011 cfm. At the same time as the people are exhaling $CO_2$, the ventilation air is bringing in outside air with a low level of $CO_2$, as diagrammed in *Figure 4.6*.

This process can be expressed in the formula:

$$VC_{space} = N + VC_{outside} \qquad (Equation\ 4\text{-}1)$$

*where* V = volume of outside air, cfm, entering the space
$C_{outside}$ = concentration, $ft^3/ft^3$, of $CO_2$ in outside air
N = volume of $CO_2$ produced by a person, cfm
$C_{space}$ = concentration, $ft^3/ft^3$, of $CO_2$ in exhaust air

**Figure 4.6** Addition of Carbon Dioxide in an Occupied Space

For the movie theatre example (the same as the hotel assembly-room) the required ventilation rate is 15 cfm per person. Inserting the values for V and N produces:

$VC_{space} = N + VC_{outside}$

$15 \text{ cfm} \cdot C_{space} = 0.011 \text{ cfm} + 15 \text{ cfm} \cdot C_{outside}$

$15 \text{ cfm} \cdot C_{space} - 15 \text{ cfm} \cdot C_{outside} = 0.011 \text{ cfm}$

$(15 \text{ cfm} \cdot C_{space} - 15 \text{ cfm} \cdot C_{outside})/15 \text{ cfm} = 0.011 \text{ cfm}/15 \text{ cfm}$

$C_{outside} - C_{space} = 0.011/15 \text{ (ft}^3/\text{ft}^3)$

$C_{outside} - C_{space} = 0.000733 \text{ (ft}^3/\text{ft}^3)$

This is about 700 parts per million of $CO_2$ in the exhaust air

Note that this calculation is based on the ventilation for one person and the $CO_2$ produced by one person. The result is the same, regardless of how many people are in the space, since everything is proportional.

The outside $CO_2$ is typically in the range of 350 to 400 parts per million, ppm, so the incoming $CO_2$ level is raised by the $CO_2$ from the occupants:

350 + 700 = 1050 ppm.

In polluted cities, the $CO_2$ level might be much higher at, say, 650 ppm, in which case the inside level will be

650 + 700 = 1350 ppm

for the same ventilation rate.

In our theatre, we can install a $CO_2$ sensor to measure the $CO_2$ level, and connect it to a controller to open the outside air dampers to maintain the $CO_2$ level at no higher than 1000 ppm. In this way the outside air provided matches the requirements of the people present. If the outside $CO_2$ concentration is above 300 ppm, then our controller, set at 1000 ppm, will cause over-ventilation rather than under-ventilation.

In this process $CO_2$ is used as a **surrogate** indicator for the number of people present.

The use of $CO_2$ control works really well in a densely populated space served by a dedicated system. It works poorly in a building with a very variable and low population.

This calculation assumes a perfect world. As we all know, this is a false assumption. The main assumptions are:

**Perfect mixing.** Mixing is usually quite good but some ventilation air may not reach the occupied space.
**Steady state.** It will take from 15 minutes to several hours for the $CO_2$ concentration to become really steady. The length of time depends on the volume of space per person. In densely populated spaces, steady state can be reached quite quickly, but in low population density areas, it can take hours.
**An even distribution of people in the space.** If people are clumped together then the level will be higher in their area and lower in the less densely occupied parts of the space.

This simple use of carbon dioxide as a surrogate cannot be used under the requirements of Standard 62.1-2004, due to the $cfm/ft^2$ ventilation requirement for the space. More sophisticated methods are possible for use under the requirements of Standard 62.1-2004, but they are beyond the scope of this course.

## The Next Step

Having introduced the ideas of: Air-conditioning zones in Chapter 2; Thermal comfort in Chapter 3; and Indoor air quality and ventilation rates in this Chapter, we will go on in Chapter 5 to consider why air conditioning zones are required, how to choose zones and how they can be controlled.

## Summary

Chapter 4 deals with the reasons for ventilating buildings, how ventilation rates are chosen for specific situations, and the how to determine and maintain good indoor air quality, IAQ.

### 4.1 Introduction

The maintenance of good indoor air quality (IAQ) is one of the major objectives of air-conditioning systems, because IAQ problems are a significant threat to health and productivity. The primary factors that influence and degrade IAQ are particles, gases, and vapors in the air.

### 4.2 Air Pollutants and Contaminants

Air pollutants and contaminants are unwanted airborne constituents that may reduce the acceptability of air. Some contaminants are brought into the conditioned space from outside, and some are generated within the space itself.

## 4.3 Indoor Air Quality Effects on Health and Comfort

Contaminants can be classified based on their effects: fatal in the short term, carcinogenic (cancer causing substances), health threatening, and annoying, with an impact on productivity and sense of well-being

## 4.4 Controlling Indoor Air Quality

Maintaining acceptable IAQ depends on the judicious use of three methods: source control, filtration, and dilution. This section also included a more detailed discussion on source control, and on filtration.

## 4.5 ASHRAE Standard 62 Ventilation for Acceptable Indoor Air Quality

ANSI/ASHRAE Standard 62, *Ventilation for Acceptable Indoor Air Quality*[1] was published in 1971, 1981 and again fully revised in 1989. The complete revisions made it easy to reference in Building Codes. In many jurisdictions, earlier versions of the standard will remain the legal requirement for many years

Since 1997, to align with the ANSI "continuous maintenance" process, the Standard is updated a bit at a time and is not required to be a consistent, whole document. Standard 62.1-2004 applies to "all indoor or enclosed spaces that people may occupy" with the provision that additional requirements may be necessary for laboratory, industrial, and other spaces.

We introduced the idea of the ventilation rate procedure, and the formula

$$VC_{space} = N + VC_{outside}$$

# Bibliography

1. ASHRAE/ANSI Standard 62.1-2004 Ventilation For Acceptable Indoor Air Quality
2. ASHRAE/ANSI Standard 62.2-2004 Ventilation and Acceptable Indoor Air Quality in Low Rise Residential Buildings
3. Industrial Ventilation published by the American Conference of Governmental Industrial Hygienists 23rd edition 1998
4. ASHRAE Standard 52.1-1992 Gravimetric and Dust Spot Procedures for Testing Air Cleaning Devices used in General Ventilation for Removing Particulate Matter
5. ASHRAE 52.2-1999 Method for Testing General Ventilation Air-Cleaning Devices for the Removal Efficiency by Particle Size

# Chapter 5
# Zones

## Contents of Chapter 5

Study Objectives of Chapter 5
5.1 Introduction
5.2 What is a Zone?
5.3 Zoning Design
5.4 Controlling the Zone
The Next Step
Summary

## Study Objectives of Chapter 5

We have talked, in a general way, about spaces and zones earlier in Chapter 2, section 2.4. In this chapter we will go into detail about the reasons for choosing zones, economic considerations and how zone controls operate. After studying the chapter, you should be able to:

Define a space and give examples of spaces.
Define a zone and give examples of zones.
List a number of reasons for zoning a building and give examples of the reasons.
Make logical choices about where to locate a thermostat.

## 5.1 Introduction

In Chapter 2, we discussed the fact that spaces have different users and different requirements, and in Chapter 4 we discussed issues of thermal comfort. To maximize thermal comfort, systems can be designed to provide independent control in the different spaces, based on their users and requirements. Each space, or group of spaces, that has an independent control is called a "**zone**."

In this chapter, we consider what constitutes a zone, the factors that influence zone choices, and the issues concerning location of the zone thermostat.

## 5.2 What is a Zone?

We have introduced and used the words "space" and "zone" in previous chapters.

To recap, a **"space"** is a part of a building that is not necessarily separated by walls and floors. A space can be large, like an aircraft hanger, or small, like a personal office.

A **"zone"** is a part of a building whose HVAC system is controlled by a single sensor. The single sensor is usually, but not always, a thermostat. Either directly or indirectly, a **thermostat** controls the temperature at its location.

A zone may include several spaces, such as a row of offices whose temperature is controlled by one thermostat in one of the offices. On the other hand a zone may be a part of a space, such as the area by the window in a large open area office.

The zone may be supplied by its own, separate HVAC system, or the zone may be supplied from a central system that has a separate control for each zone.

Some examples of spaces and zones are shown in *Figure 5.1*.

Having established the meaning of a zone let us now consider the various reasons for having zones in a building's HVAC system.

| Space | Zones | Reason for zones |
|---|---|---|
| A theatre used for live performance | 1. Audience seating | The audience area requires cooling and high ventilation when the audience is present. |
| | 2. Stage | The stage requires low ventilation and low cooling until all the lights are turned on, and then high cooling is required. |
| Indoor ice rink | 1. Spectators | Spectators need ventilation and warmth. |
| | 2. Ice sheet | The ice sheet needs low air speeds and low temperature to minimize melting. |
| | 3. Space above | The space above the spectators and ice may need moisture removal to prevent fogging |
| Deep office | 1. By the windows | People by the window may be affected by the heat load from the sun and by the cool window in winter, external factors. |
| | 2. Interior area | The interior zone load will change due to the occupants, lights, and any equipment – a cooling load all year. |
| Large church or mosque | 1. Within 6 feet of the floor | The occupied zone is within 6 feet of the floor and needs to be comfortably warm or cool for congregation. |
| | 2. Above six feet | The space above does not need to be conditioned for the congregation |
| Airport | 1. Lobby | This is a huge space with a variety of uses, and extremely variable occupancy and loads. |
| | 2. Security | |
| | 3. Retail outlets | |
| | 4. Check-in | Each zone requires its own conditions. |

**Figure 5.1** Examples of Spaces and Zones

## 5.3 Zoning Design

There are several types of zones. These zones are differentiated based on what is to be controlled, and the variability of what is to be controlled. The most common control parameters are: thermal (temperature), humidity, ventilation, operating periods, freeze protection, pressure and importance.

The most common reason for needing zones is variation in thermal loads. Consider the simple building floor plan shown in *Figure 5.2*. Let us assume it has the following characteristics:

Well-insulated
A multi-story building, identical plan on every floor
Provided with significant areas of window for all exterior spaces
Low loads due to people and equipment in all spaces
Located in the northern hemisphere

In this example, we will first consider the perimeter zone requirements on intermediate floors due to changes in thermal loads. These changes can occur because of the movement of the sun around the building during the course of a sunny day. These changes in thermal loads take place because the spaces receive solar heat from the sun, called **solar gain**.

The designer's objective is to use zones to keep all spaces at the set-point temperature. The **set-point temperature** is the temperature that the thermostat is set to maintain.

Early in the morning, the sun rises in the east. It shines on the easterly walls and through the east windows into spaces NE and SE. Relative to the rest of the building, these spaces, NE and SE, need more cooling to stay at the set-point temperature.

As the morning progresses towards midday, the sun moves around to the south so that the SE, S and SW spaces receive solar gain. However, the solar heat load for the NE space has dropped, since the sun has moved around the building.

As the afternoon progresses, the sun moves around to the west to provide solar gain to spaces SW and NW.

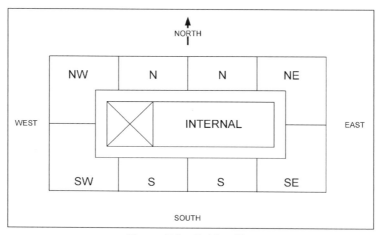

**Figure 5.2** Building Plan

## Zoning Design Considerations

While most of the spaces have been experiencing a period of solar gain, the two N spaces have had no direct solar gain. Thus, the load in the two N spaces is only dependent on the outside temperature and internal loads, like lights. These two factors are approximately the same for each space. Therefore these two N spaces could share a common thermostat to control their temperature and it would not matter whether the thermostat was located in one space or the other. These two spaces would then be a single zone, sharing a single thermostat for the temperature control of the two spaces.

The two S spaces have similar thermal conditions with high solar gain through the middle of the day. Both of the two S spaces could also share a thermostat, since they have similar solar and other loads.

The remaining spaces: NE, SE, SW, and NW, all have different solar gains at different times. In order to maintain the set-point temperature, they would each need their own thermostat.

Thus, if we wanted to deal with the solar gain variability in each of these eight spaces, we would need six zones. Note that this discussion is considering zoning on the basis of only solar loads.

In real life there may not be enough funds allocated for six zones. Thus, the designer might combine the two N spaces with the NE space; on the basis that a little overheating in NE space in the early morning would be acceptable. Then the choice is between N and NE spaces for the thermostat location. Since it is generally better to keep the majority happy, the designer would choose to put the thermostat in an N space. However, if the designer knew that the NE space was going to be allocated to an important person, the choice could be to put the thermostat in the NE space!

In a similar way the two S spaces and the SE space could be combined, since they all experience the midday solar gain. Lastly the SW and NW spaces could be combined, since they both experience the high solar gain of the late afternoon.

In this way, the six zones could be reduced to three. The effect would be to have considerable loss of temperature control performance, but there would also be a coincident reduction in the installation cost.

The balance between performance and cost is a constant challenge for the designer. Too few zones could lead to unacceptable performance and potential liability, while excessive zoning increases costs and maintenance requirements.

## Interior and Roof Zones

The discussion so far has ignored both the internal zone and the effect of the roof. The internal zones on intermediate floors are surrounded by conditioned spaces. As a result, they never need heating, are not affected by solar gain and need cooling when occupied all year. In a cool climate this can often create a situation where all exterior zones require heating but the interior zones still require cooling. The different behavior of interior zones can be dealt with by putting them on a separate system.

The top floor perimeter zones are also different from the intermediate floor zones since they have the added summer roof solar gain and the winter heat loss. On the top floor, interior zones are also affected by solar gain and winter heat loss. As a result the top floor design needs special consideration with additional cooling and heating abilities.

Choosing zones is always a cost/benefit trade-off issue. In an ideal world, every occupant would have control of their own part of the space. In practice the cost is generally not warranted. As a result the designer has to go through a selection process, like we did in this example, to decide which spaces in a building can be combined. In our example, we only considered solar gain, but in a real building the designer must consider all relevant factors. Common factors are outlined below:

### Thermal Variation

*Solar gain*. As shown in the example, solar gain through windows can create a significant difference in cooling load, or the need for heating, at varying times of the day according to window orientation.

*Wall or roof heat gains or heat losses*. The spaces under the roof in a multi-floor building will experience more heat gain in summer, or heat loss in winter, than spaces on the lower floors.

*Occupancy*. The use of spaces and the importance of maintaining good temperature control will influence how critical zoning is.

*Equipment and associated heat loads*. Equipment that gives off significant heat may require a separate zone in order to maintain a reasonable temperature for the occupants. For example, a row of private offices may have worked well as a single zone, but the addition of a personal computer and a server in one of those offices would make it very warm compared to the other offices. This office could require separate zone design.

*Freeze protection in cold climates*. In a cold climate, the perimeter walls and roof lose heat to the outside. Therefore, it is often advantageous to designate the perimeter spaces as separate heating zones from those in the core of the building.

### Ventilation with Outside Air

*Occupancy by people*. In a typical office building, the population density is relatively low. However, conference rooms have a fairly high potential population density and therefore, a very variable, and not continuous, ventilation load. Therefore, conference rooms are often treated as different zones for ventilation **and** for time of operation, compared to the offices in a building.

*Exhausts from washrooms*. As noted in Chapter 4, washrooms may be treated as a separate zone and provided only with exhaust. The exhausted air may be made up of air from the surrounding spaces.

*Exhausts from equipment and fume hoods*. Often, equipment is required to operate continuously, although the majority of the building is only occupied during working hours, Monday to Friday. In these cases, it may be advantageous to treat the spaces with continuous exhaust as a separate zone or even service them from a separate system.

### Time of Operation

*Timed*. In many buildings, the time of operation of spaces differs. For example, an office building might have several floors occupied by tenants who are happy with full service only during working hours from Monday to Friday. One floor could be occupied by a weather forecasting organization that required full operation 24 hours-a-day seven days a week. In this case it might be advantageous to have the building zoned to only provide service when and where needed.

*On demand – manual control or manual start for timed run.* In many buildings there are spaces that are only used on occasion. They may be designed as separate zones, which are switched on when needed. The activation can be by means of an occupancy sensor, or by a manual start switch in the space, which runs the zone for a predetermined time. For example a low-use lecture theatre in a university building might be provided with a push button start that would energize the controls to run the space air conditioning for two hours before switching off.

### Humidity

*High humidity in hot humid climates for mold protection.* In hot, humid climates, the moisture can infiltrate into the building through leaks in the walls, doors and windows. This can cause the building contents to mold unless dehumidification is activated.

Humidity sensors can be installed in individual representative zones that will measure relative humidity. If these sensors detect excess humidity in these zones, they can trigger the system to provide system wide dehumidification. The control system can be designed to provide dehumidification without ventilation during unoccupied hours.

*Museum and art gallery requirements for good humidity control.* High quality museums and art galleries have to maintain accurate control of the humidity in the storage and exhibit areas. This humidity control is generally not required in other spaces like offices, restaurants, merchandising and lobby areas. Therefore museum and art gallery often have at least two systems, to provide the collections with the required humidity control.

### Pressure

Air flows from places at a higher pressure to places at a lower pressure.

A difference in pressure can be used to control the movement of airborne contaminants in the building. For example, in a hospital, the tuberculosis (TB) patient rooms can be kept at a negative pressure compared to surrounding areas, to ensure that no TB germs, known as bacilli, migrate into surrounding areas.

In a similar way, kitchens, smoking rooms, and toilets are kept negative to contain the smells by exhausting more air than is supplied to the spaces. Conversely, a photographic processing laboratory is kept at a positive pressure to minimize the entry of dust.

### Zoning Problems

One recurring problem with zoning is changes in building use after the design has been completed. If there are likely to be significant changes in layout or use, then the designer should choose a system and select zones that will make zone modification as economical and easy as practical.

Having reviewed the reasons for choosing to zone a building, let's consider the control of the zone.

## 5.4 Controlling the Zone

The most common zone control device is the thermostat. It should be placed where it is most representative of the occupants' thermal experience. A thermostat is usually mounted on the wall. It is designed to keep a constant temperature where it is, but it has no intelligence; it does not know what is going on around it. The following are some of the issues to be aware of when choosing the thermostat location.

- Mounting the thermostat in a location where the sun can shine on it will cause it to overcool the zone when the sun shines on it. The sun provides considerable radiant heat to the thermostat. The thermostat interprets the radiant heat as though the whole location had grown too warm, and it will signal the air conditioning system that it requires a lower air temperature. As a result, the occupants will be cold, and cooling expenses will escalate.
- In many hotels, the thermostat is mounted by the door to the meeting room. If the door is left open, a cold or warm draft from the corridor can significantly, and randomly, influence the thermostat.
- In some conference or assembly rooms, the thermostat is mounted above lighting dimmer switches. These switches produce heat that rises up into the thermostat. This makes the thermostat think that the room is warmer than it actually is. If the dimmers are left alone and their output is constant, the thermostat can be set at a set point that allows for the heating from the dimmers. Unfortunately the dimmers heat output changes if the dimmer setting is adjusted, so adjusting the lighting level will alter the thermostat performance.
- Mounting a thermostat on an outside wall can also cause problems. If the wall becomes warm due to the sun shining on it, the thermostat will lower the air temperature to compensate. This offsets the increased radiant temperature of the wall on the occupants, but usually the effect is far too much and the room becomes cool for the occupants. In a similar way, in the winter the wall becomes cool and a cool draft will move down the wall over the thermostat, causing it to raise the air temperature to compensate.
- There are times when heat from equipment can offset the thermostat. A computer mounted on a desk under a thermostat can easily generate enough heat to cause the thermostat to lower the air temperature. If the computer is only turned on periodically, (perhaps to drive a printer,) this offset will occur at apparently random times, creating a difficult problem for the maintenance staff to resolve.
- If the thermostat is mounted where it is directly affected by the heating or the cooling of the space, it will likely not maintain comfortable conditions. For example, let us imagine that the air-conditioning system air-supply blows directly onto the thermostat. In the heating mode, the thermostat will warm up quickly when the hot air stream blows over it. Therefore, it will quickly determine that the room is warm enough and turn off the heat. The result will be rapid cycling of the thermostat and the room will be kept cooler than the set-point temperature. Conversely, when in the cooling mode, the thermostat will be quickly cooled and will cycle rapidly, keeping the room warmer than the set-point temperature.

If the system has been adjusted to work satisfactorily during the heating season, then when the system changes over to cooling, the thermostat will keep the zone warmer than it did when in the heating mode. Complaints will result and the thermostat will get adjusted to satisfactory operation in the cooling mode. When the season changes, the shift will reverse and

readjustment will be required once more. This is the sort of regular seasonal problem that occurs in many air-conditioning systems.
- Wall-mounted thermostats generally have a cable connecting them to the rest of the control system. The hole, tubing or conduit can allow air from an adjoining space or the ceiling to blow into the thermostat, giving it a false signal.
- Lastly, mounting a thermostat near an opening window can also cause random air temperature variations as outside air blows, or does not blow, over the thermostat.

### Humidity

While this discussion has been all about thermostats and poor temperature control, the issues are very similar for humidity, which is controlled by **humidistats**. The result of failing to consider placement of the humidistat will be poor humidity control. Remember, as we discussed in Chapter 2, section 2.2.1, if the temperature rises, then relative humidity drops and conversely, if the temperature falls then the humidity rises.

## The Next Step

Having considered the issues around zones, we are now going to consider typical systems that provide zone control. In Chapter 6 we will be considering single zone systems and in Chapter 7, systems with many zones.

## Summary

### 5.2 What is a Zone?

A zone is a part of a building whose HVAC system is controlled by a single sensor. The single sensor is usually, but not always, a thermostat. Either directly or indirectly, a thermostat controls the temperature at its location.

### 5.3 Zoning Design

Zones are chosen based on what is to be controlled and the variability of what is to be controlled. The most common control parameters include: temperature, humidity, ventilation, operating periods, freeze protection, pressure, and importance.

### 5.4 Controlling the Zone

The most common zone control is the thermostat. It should be placed where it is most representative of the occupants' thermal experience. A thermostat does its best to keep a constant temperature where it is. It has no intelligence; it does not know what is going on around it. Therefore, in order to maintain a set point for the zone, the thermostat must be located away from temperature affecting sources, like drafts, windows and equipment.

Chapter 6

# Single Zone Air Handlers and Unitary Equipment

## Contents of Chapter 6

Study Objectives of Chapter 6
6.1 Introduction
6.2 Examples of Buildings with Single-zone Package Air-Conditioning Units
6.3 Air-Handling Unit Components
6.4 Refrigeration Equipment
6.5 System Performance Requirements
6.6 Rooftop Units
6.7 Split Systems
The Next Step
Summary
Bibliography

## Study Objectives of Chapter 6

After studying Chapter 6, you will be able to:

Identify the main components of a single zone air handler and describe their operation.
Describe the parameters that have to be known to choose an air-conditioning air-handling unit.
Describe how the vapor compression refrigeration cycle works.
Identify the significant issues in choosing a single-zone rooftop air-conditioning unit.
Understand the virtues of a split system.

## 6.1 Introduction

In the previous chapters we have discussed ventilation for maintaining indoor air quality, the thermal requirements for comfort, and reasons for zoning a building. In this chapter we are going to consider packaged single-zone air-conditioning equipment, examine issues of system choice and provide a general description of system control issues. We will return to controls in more depth in Chapter 11.

The single-zone air-conditioning equipment we will be discussing is the piece of equipment that was introduced in Chapter 2, *Figure 2.12*. This unit is typically referred to as the **single zone air handler, or air-handling unit, often abbreviated to AHU**. In this chapter, we will refer to it as the **air handler** or the **unit**. The air handler draws in and mixes outside air with air that is being recirculated, or returned from the building, called **return air**. Once the outside air and the return air are mixed, the unit conditions the mixed air, blows the conditioned air into the space and exhausts any excess air to outside, using the return-air fan.

Before getting into a discussion of the components of a single-zone package air-conditioning unit, we need some context as to where it fits into the whole building or site systems.

## 6.2 Examples of Buildings with Single-zone Package Air-Conditioning Units

*Figure 6.1* shows four identical single story buildings, A, B, C, and D. Each has a single-zone package air-conditioning unit (marked "**AHU**") located on the roof.

- **Building A**: This unit has only an electrical supply. This single electrical supply provides all the power for heating, cooling, humidifying, and for driving the fans.
- **Building B**: This unit has the electrical supply for cooling, humidifying, and for driving the fans, while the gas line, shown as "gas supply," provides heating.

These first two arrangements are commonly available as factory engineered, off the shelf, rooftop packages. Among these packaged units, there is a great range in size, quality, and features. The most basic provide few, if any, options. They are relatively difficult to service and have a relatively short life. At the other end of the spectrum, there are large units with walk-in service access and numerous energy-conserving options. These are designed to last as long as any indoor equipment.

As well as the total pre-packaged units, there are units, typically in larger buildings or complexes of buildings, where the heating is provided from a central service. For example, a boiler room can produce hot water that is piped around the building or buildings to provide heat. Each air-handling unit that needs heating has hot water piped to it.

- **Building C**: This unit has the electrical supply for cooling, humidifying, and for driving the fans. It also has supply and return hot-water pipes coming from a boiler room in another building. The unit contains a hot-water heating coil and control valve, which together take as much heat as needed from the hot water supply system.
- **Building D**: In the same way, there may be a central chiller plant that produces cold water at 42°F – 48°F, called **chilled water**. This chilled water is piped around the building, or buildings, to provide the air-handling units with cooling. Like the heating coil and control valve in Building C, there will be a cooling coil and control valve in each unit, to provide the cooling and dehumidification.

70　Fundamentals of HVAC

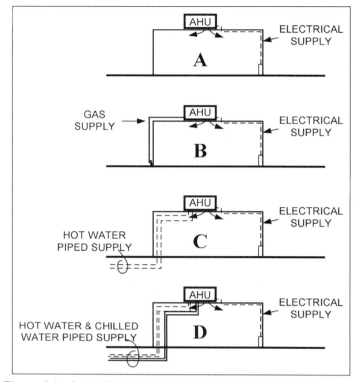

**Figure 6.1**　Single Zone Rooftop Air-Conditioning Unit, Energy Supplies

To recap, a packaged unit can require just an electrical source of power, or it may get heating in the form of a gas or hot water supply, and may get cooling from a source of chilled water. The basic operation of the unit stays the same; it is just the source of heating and cooling energy that may change.

## 6.3 Air-Handling Unit Components

You should recognize *Figure 6.2*, which was originally introduced in Chapter 2, Figure 2.12. It shows the basic air-handler unit with the economizer cycle. Some new details have been added in this diagram. In the following section, we will go through each of the components in the unit, we will discuss what each component does, and, in general terms, how each component can be controlled. This unit is typically referred to as the **single-zone air handler**.

The overall functions of the air-handler are to draw in outside air and return air, mix them, condition the mixed air, blow the conditioned air into the space, and exhaust any excess air to outside.

### Air Inlet and Mixing Section

The inlet louver and screen restrict entry into the system. The inlet louver is designed to minimize the entry of rain and snow. A very simple design for the inlet louver is shown in the diagram. Maintaining slow air-speed through the

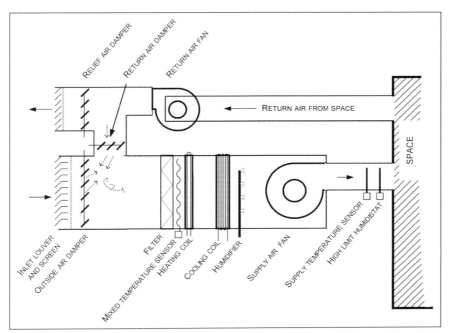

**Figure 6.2** Air-Conditioning System: Single-Zone Air Handler

louver avoids drawing rain into the system. More sophisticated, and more costly designs allow higher inlet-velocities without bringing in the rain. The screen is usually a robust galvanized-iron mesh, which restricts entry of animals, birds, insects, leaves, etc.

Once the outside air has been drawn in, it is mixed with return air. In *Figure 6.2*, a parallel blade damper is shown for both the outside air damper and the relief air damper.

$$/ / / / / /$$

These dampers direct the air streams toward each other, causing turbulence and mixing. Mixing the air streams is extremely important in very cold climates, since the outside air could freeze coils that contain water as the heating medium. A special mixing section is installed in some systems where there is very little space for the mixing to naturally occur.

It is also possible to install opposed blade dampers:

$$/ \backslash / \backslash / \backslash$$

These do a better job of accurately controlling the flow, but a rather poorer job of promoting mixing.

Some air will be exhausted directly to the outside from washrooms and other specific sources, like kitchens. The remainder will be drawn back through the return air duct by the return air fan and either used as return air, or exhausted to outside through the relief air damper. This exhausted air is called the **relief air**. The relief air plus the washroom exhaust and other specific exhaust air will approximately equal the outside air that is brought in. Thus, as the incoming outside air increases, so does the relief air. It is common, therefore, to link the outside-air damper, the return-air damper and the relief-air dampers and use a single device, called an **actuator**, to move the dampers in unison. When the system is "off," the outside-air and relief-air dampers are fully closed, and the return-air damper is fully open. The system can be started and all the air will recirculate through the return damper. As the damper actuator drives the three dampers, the outside-air and relief-air dampers open in unison as the return-air damper closes.

### Mixed Temperature Sensor

Generally, the control system needs to know the temperature of the mixed air for temperature control. A mixed-temperature sensor can be strung across the air stream to obtain an average temperature. If mixing is poor, then the average temperature will be incorrect. To maximize mixing before the temperature is measured, the mixed temperature sensor is usually installed downstream of the filter.

When the plant starts up, the return air flows through the return damper and over the mixed temperature sensor. Because there is no outside air in the flow, the mixed-air temperature is equal to the return-air temperature. The dampers open, and outside air is brought into the system, upstream of the mixed-air sensor. If the outside temperature is higher than the return temperature, as the proportion of outside air is increased, the mixed-air temperature will rise. Conversely, if it is cold outside, as the proportion of outside air is increased, the mixed-air temperature will drop. In this situation, it is common to set the control system to provide a mixed-air temperature somewhere between 55 and 60°F. The control system can simply adjust the position of the dampers to maintain the set mixed temperature.

For example, consider a system with a required mixed temperature of 55°F and return temperature of 73°F. When the outside temperature is 55°F, 100% outside air will provide the required 55°F. When the outside air temperature is below 55°F, the required mixed temperature of 55°F can be achieved by mixing outside air and return air. As the outside temperature drops, the percentage required to maintain 55°F will decrease. If the return temperature is 73°F, at 37°F there will be 50% outside air, and at 1°F, 20% outside air.

If the building's ventilation requirements are for a minimum of 20% outside air, then any outside temperature below 1°F will cause the mixed temperature to drop below 55°F. In this situation, the mixed air will be cooler than 55°F and will have to be heated to maintain 55°F. The mixed-air temperature-sensor will register a temperature below 55°F. The heating coil will then turn "on" to provide enough heat to raise the supply-air temperature (as measured by the supply-temperature sensor) to 55°F.

Now let us consider what happens when the outside-air temperature rises above 55°F. Up to 73°F, the temperature of the outside air will be lower than the return air, so it would seem best to use 100% outside air until the outside temperature reaches 73°F. In practice, this is not always true, because the moisture content of the outside air will influence the decision. In a very damp climate, the changeover will be set much lower than 73°F, since the enthalpy of the moist, outside air will be much higher than the dryer return air, at 73°F. Above the pre-determined changeover temperature, the dampers revert to the minimum ventilation rate, 20% outside air in this example.

The last few paragraphs have discussed the how the system is controlled, called the control operation. These control operations can be summarized in the following point form, often called the **control logic**:

When system off, the outside air and relief air dampers fully closed, return air dampers fully open.
When system starts, if outside temperature above 70°F, adjust dampers to provide x cubic feet per minute (cfm) of outside air.
When system starts, if outside temperature below 70°F, modulate dampers to maintain 55°F mixed temperature with a minimum of x cfm of outside air.

The requirement for a minimum volume of outside air means that the controller must have a way of measuring the outside air volume. This can be achieved in a number of ways that are explained in the ASHRAE Course *Fundamentals of Air System Design*[1].

The preceding text has talked about air volumes without getting into specific numbers. Note that the weight (mass if you leave earth) of outside air entering the building must equal the weight of air that leaves the building. The volume of air that is entering and leaving will usually be different, since the volume increases with increasing temperature. For example:

82 lb/min, 1000 cfm of outside air, at 25°F, enters a building.

It is heated, and leaves the building as

82 lb/min, 1100 cfm at 75°F (10% greater volume, same weight)

## *Filter*

All packaged units include as least minimal filters. Often it is beneficial to specify better filters, as we discussed in Chapter 4.4.

## *Heating Coil*

Some systems require very high proportions, or even 100% outside air. In most climates this will necessitate installing a heating coil to raise the mixed air temperature. The heat for the heating coil can be provided by electricity, gas, water or steam.

The electric coil is the simplest choice, but the cost of electricity often makes it an uneconomic one.

A gas-fired heater often has the advantage of lower fuel cost, but control can be an issue. Inexpensive gas heaters are "on-off" or "high-low-off" rather than fully modulating. As a result, the output temperature has step changes. If the unit runs continuously with the heat turning on and off, then the supply temperature will go up and down with the heater cycle and occupants may experience a draft.

Hot water coils are the most controllable, but there is a possibility that they will freeze in cold weather. If below-freezing temperatures are common, then it is wise to take precautions against coil freezing. Many designers will, therefore, include a low-temperature alarm and arrange the controls to keep the coil warm or hot, when the unit is off during cold weather.

This is one of the times when the designer needs to take precautions against the consequences of the failure of a component. If, for example, the damper linkage fails, the unit may be "off," with the outside dampers partially open to the freezing weather. The consequence, a frozen coil, is serious since it will take time to get it repaired or replaced.

### *Cooling Coil*

Cooling is usually achieved with a coil cooled by cold water, or a refrigerant. The cold water is normally between 42°F and 48°F. There are numerous refrigerants that can be used, and we will discus the refrigerant cycle and how it works in the next section. Whether using chilled water or a refrigerant, the coil will normally be cooler than the dew point of the air and thus condensation will occur on the coil. This condensation will run down the coil fins to drain away.

With refrigeration coils in packaged systems, there is limited choice in the dehumidification capacity of the coil.

### *Humidifier*

A humidifier is a device for adding moisture to the air. The humidifier can either inject a water-spray or steam into the air.

The water-spray consists of very fine droplets, which evaporate into the air. The supply of water must be from a **potable** source, fit for human consumption. If impurities have not been removed by reverse osmosis or some other method, the solids will form a very fine dust as the water droplets evaporate. This dust may, or may not, be acceptable.

The alternative is to inject steam into the air stream. Again, the steam must be potable.

The humidifier will normally be controlled by a humidistat, which is mounted in the space or in the return airflow from the space. Excessive operation of the humidifier could cause condensation on the duct surface and result in water dripping out of the duct. To avoid this possibility, a high humidity sensor is often installed in the duct, just downstream from the unit. In addition, one might not want the humidifier to run when the cooling coil is in operation.

The unit control logic will then be:

Humidifier off when unit off
Humidifier off when cooling in operation
Humidifier controlled by space humidistat when unit in operation
Humidifier to shut down until manually reset if high limit humidity sensor
    operates

## Fan

The fan provides the energy to drive the air through the system. There are two basic types of fan, the **centrifugal**, and the **axial**.

Within the **centrifugal fan**, air enters a cylindrical set of rotating blades and is centrifuged, thrust radially outwards, into a scroll casing. This fan is a very popular choice due to its ability to generate substantial pressure without excessive noise.

The other type of fan is the **axial fan**, where the air passes through a rotating set of blades, like an aircraft propeller, which pushes the air along. This is a simpler, straight-through design that works really well in situations that require high volumes at a low pressure-drop. When this type of fan is made for really low pressure-drops, wide pressed-sheet-metal blades are used and it is called a propeller fan.

## Return Fan

A return fan is usually included on larger systems, unless there is some other exhaust system to control building pressure. If there is no return fan, the building will have a pressure that is a bit above ambient (outside). In a hot, humid climate, this is beneficial since it minimizes the infiltration of outside air into the building, where it could cause condensation and mildew. In cold climates, the excess pressure above ambient can cause leakage of moist air into the wall, where it freezes and causes serious damage.

Having briefly reviewed the unit components, we are going to take time to consider the refrigeration cycle and its operation.

## 6.4 Refrigeration Equipment

Heat naturally flows from warmer places to cooler places. Refrigeration equipment is used to transfer heat from a cooler place to a warmer place. In the domestic refrigerator, the refrigeration equipment absorbs heat from inside the refrigerator and discharges heat into the house. On a much larger scale, refrigeration machines are used to chill water that is then pumped around buildings to provide cooling in air-conditioning systems. The heat removed from the water is expelled into the atmosphere through a hot, air-cooled coil, or by evaporating water in a cooling tower.

The domestic refrigerator and most other refrigeration systems use the same basic process of vapor compression and expansion. An alternative process, adsorption, is used but we are not covering it in this course. The vapor compression refrigeration system comprises four components: compressor, condenser, expansion valve, and evaporator. *Figure 6.3* shows the arrangement.

*Compressor*—which compresses refrigerant vapor to a high pressure, making it hot in the process.

*Condenser*—in which air or water cooling reduces the temperature of the refrigerant sufficiently to cause it to condense into liquid refrigerant and give up its latent heat of evaporation. Latent heat of evaporation is the heat required to convert a liquid to a vapor at a particular temperature and pressure and is the heat released when a vapor condenses at a particular temperature and pressure.

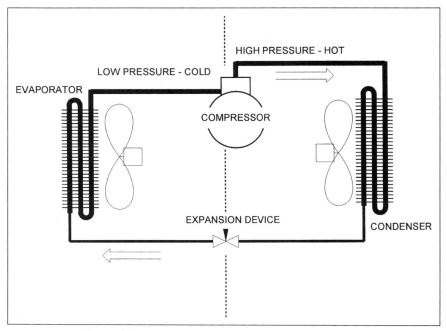

**Figure 6.3** Basic Vapor Compression Refrigeration Cycle

*Expansion valve*—which allows a controlled amount of the liquid refrigerant to flow through into the low-pressure section of the circuit.

*Evaporator*—in which air or water heats the liquid refrigerant so that it evaporates (boils) back into a vapor as it absorbs its latent heat of evaporation.

As the refrigerant flows round and round the circuit, it picks up enthalpy, heat, at the evaporator and more heat as it is compressed in the compressor. The sum of the evaporator and compressor enthalpy is rejected from the condenser. The system effectiveness is higher, the greater the ratio of evaporator enthalpy to compressor enthalpy. One wants the most heat transferred for the least compressor work. The enthalpy flow into and out of the refrigerant is shown in the *Figure 6.4*.

In a very small, simple system, such as the domestic refrigerator, the expansion device is a length of very small-bore tube that restricts the refrigerant liquid flow from the high-pressure side to the low-pressure side. A thermostat in the refrigerator turns the compressor "on" when cooling is required, and "off" again when the inside of the refrigerator is cool enough.

Moving up in size from the domestic refrigerator to the window air conditioner, *Figure 6.5* shows the refrigeration circuit with a box around it. The evaporator fan draws room air over the evaporator coil to cool it. The condenser is outside and the condenser fan draws outside air over the condenser coil to reject heat into the outside air.

The evaporator coil is designed to operate cool enough to produce some condensation on the coil. This condensate water is piped through to the outside and may just drip out of the unit or be evaporated in the condenser airflow.

Single Zone Air Handlers and Unitary Equipment 77

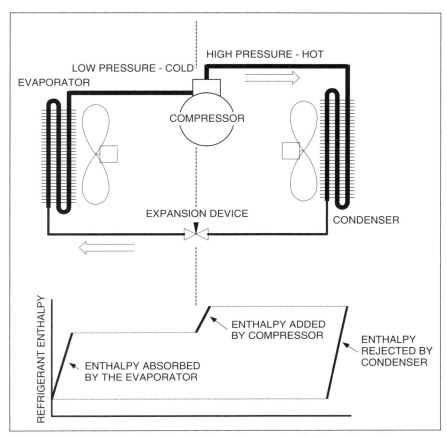

**Figure 6.4** Enthalpy Flow in Vapor Compression Refrigeration Cycle

The capacity of the unit is highest when the inside and outside temperatures are close to each other. As the outside temperature rises, so the capacity of the unit falls. It is therefore very important to know the anticipated maximum temperature at which the unit is to perform.

The refrigerator and the window air conditioner have air flowing across both the evaporator and condenser to achieve heat transfer. Many systems use water as an intermediate heat-transfer medium. The evaporator coil can be in a water-filled shell to produce chilled water. This chilled water can then be piped around the building, or even from building to building, to provide cooling as and where it is needed.

This central water-chilling plant can consist of one or more chillers that are sequenced to match their capacity with the load. In this way the noisy refrigeration equipment can be separated from occupied areas, and maintenance does not take place in occupied areas.

Water can also be used on the condenser side of the refrigeration system. Here the condenser heats the water, which is generally then pumped to one or more cooling towers. A **cooling tower** is a piece of equipment for cooling water by evaporation. The warmed condenser water enters at the top through a series of nozzles, which spread the water over an array of wooden or plastic surfaces.

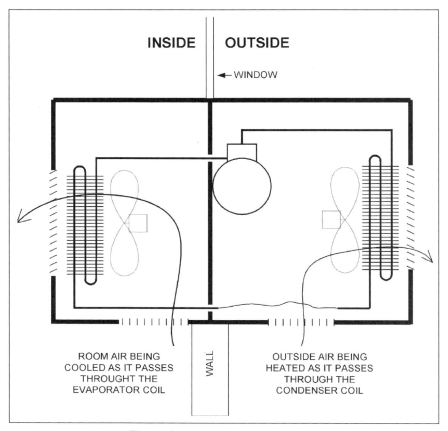

**Figure 6.5** Window Air Conditioner

Most cooling towers also have a fan to force air through the surfaces, causing some of the water to evaporate and cool the remaining water. The cooled water flows down into a sump, to be pumped back through the condenser.

### Heat Pump

The previous discussion is focused on pumping heat from a cooled space and rejecting heat to outside. There are times when the reverse process is valuable. If the outside temperature is not too cold, one could install a window air conditioner back-to-front. Then, it would cool outside and warm inside. The total heat rejected to the inside would be the sum of the electrical energy put into the compressor, plus heat absorbed from the outside air. It would be pumping the heat into the space – hence we call it a **heat pump**. In milder climates, a heat pump can obtain useful heat from the ambient air.

In practice, one does not take out the window air conditioner and install it the other-way-round for heating, since the reversal can be achieved with a special valve in the refrigeration circuit. *Figure 6.6* shows the heat pump circuit. It has been drawn slightly differently from the previous two figures, but

Single Zone Air Handlers and Unitary Equipment    79

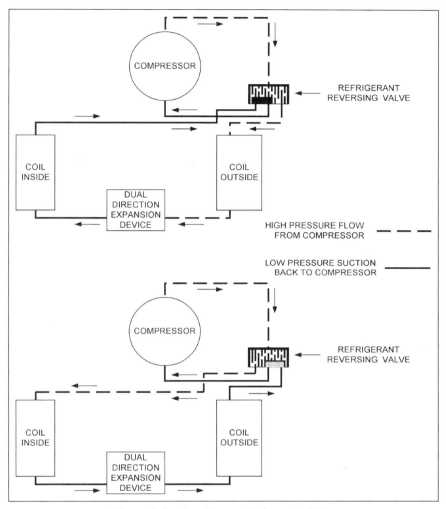

**Figure 6.6**   Heat Pump with Reversing Valve

the circuit is the same, evaporator, compressor, condenser, and expansion device. In the upper diagram the refrigerant is flowing, as in previous diagrams, and heat is being 'pumped' from the inside coil to be rejected by the outside coil. In the lower diagram the reversing valve has been switched to reverse the flow of refrigerant in the inside and outside coils. Heat in now absorbed from outside and rejected by the inside coil, heating the inside.

The performance of the air-to-air heat pump drops as the temperature difference increases, so they are not very effective with an outside air temperature below freezing.

Another source of heat, or sink for waste heat, is the ground. In many places, one can lay coils of pipe in the ground, in trenches or in vertical boreholes, and circulate water. The water will be heated by the surrounding soil, if it is cold, and cooled by the surrounding soil if it is hot. In the example, shown in *Figure 6.6*, the heat pump has a ground water heated/cooled coil and a cooled/heated air coil for the building. *Figure 6.6* shows the circuit, including the reversing valve operation.

Refrigeration is a very important part of the air-conditioning industry. The ASHRAE Course, *Fundamentals of Refrigeration*[2] will teach you about the systems, components, system control and cooling loads.

## 6.5 System Performance Requirements

Before choosing a system, you need an understanding of the types of loads you want the system to manage. Typically, the summer cooling-loads will be the main determinant of the choice of unit. The heating loads are usually easily dealt with by choosing a suitable heater to go with the chosen unit. The summer loads, though, will be dependent on several, somewhat interrelated factors:

*Outside summer design temperature.* This affects the cooling load in three ways:

> *Interior load*—The interior load is calculated using the outside temperature plus solar heat gain acquired due to heat transfer through the fabric of the building.
> *Outside air temperature*—The load from the outside air temperature will also partly determine the cooling load of the outside air that is being brought into the building for ventilation.
> *Effectiveness of the refrigeration system*—If the refrigeration system is air-cooled, the outside temperature will influence the effectiveness of the refrigeration system.

*Outside summer design humidity.* The outside design humidity will be a factor in the ventilation air load and the removal of moisture from any air that leaks into the building. Cooling tower performance is also directly affected by the humidity; performance falls as humidity rises.

**Inside summer design temperature and humidity.** The warmer and damper the inside is allowed to be, the smaller the difference between inside and outside, hence the lower the load on the system. This is particularly important when you are making system choices.

Slight under-sizing, which is cheaper to buy, means that occasionally the design temperatures will be exceeded. However, when the unit is slightly under-sized, it will be running nearer full load for more of the time. Depending on the situation, this may be the most economical choice.

*Inside summer generation of heat and moisture.* These will be added to the building loads to establish the total loads on the system.

*Summer ventilation requirements.* This is the ventilation for people, typically based on ASHRAE Standard 62.1, 2004 plus any additional ventilation for specific equipment. The higher the ventilation requirements, the greater the load due to cooling and dehumidifying the outside air that is brought in.

Once these basic criteria are established, load calculation can be done. Depending on the situation, summer cooling and winter heating loads may be estimated with fairly simple hand calculation methods for the peak-load summer-cooling and for the peak-load winter-heating. In other cases, an hour-by-hour computer simulation of the building may be done, in order to assess peak-load and intermediate-load performance.

The following example illustrates some of the issues for system performance. **A building has the following conditions**:

The design room condition is 75°F and 50% relative humidity.
The outside design condition is 95°F and 40% relative humidity.
The sensible heat load is 200,000 Btu/h. **Sensible heat** is heat that causes change in temperature.
The moisture heat load, or "Latent heat" is 20,000 Btu/h. **Latent heat** is the energy that is absorbed by water which causes the water to evaporate.

To calculate the loads, first divide the latent load by the sensible load. This provides us with the percentage of sensible heat that must be removed from the system.

For example, if the latent load is 20,000 Btu per hour (Btu/h) and the sensible load is 200,000 Btu/h, the ratio would be 1/10. With an all-air system, the air supply must be at a temperature and moisture content that requires 10 times as much sensible heat as latent heat to reach room temperature. We can plot a line on which the air supply must be to meet the design room condition.

You can easily plot this on the psychrometric chart as is shown in *Figure 6.7*.

First, note the enthalpy of the air at the desired room condition.
Draw a vertical line downward from the room condition.
Mark on the line where the enthalpy line is 1 Btu/lb less than room conditions.
From this point, draw a horizontal line to the left. Mark off where the enthalpy is 10 Btu/lb less.
Draw a line from here to the room condition.

This illustrates that, for the supply air to meet the designed room condition, it must be supplied at some point on this line. If it is supplied close to the designed room-condition the volume will have to be large.

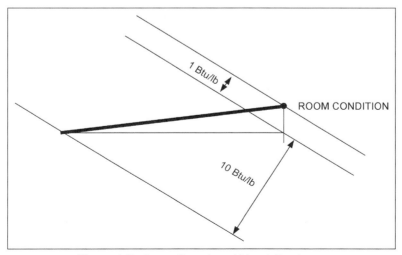

**Figure 6.7** Space, Outside and Mixed Conditions

82    Fundamentals of HVAC

The calculation of heating loads is relatively straightforward, but cooling-load calculation is more challenging, due to the movement of the sun and changing loads throughout the day. Calculating heating and cooling loads is the subject of the ASHRAE Course *Fundamentals of Heating and Cooling Loads*[3].

### Decision Factors for Choosing Units

When choosing equipment, several factors must be balanced.

> **The initial cost to purchase and install versus the ongoing cost of operation and maintenance**. Most heating and cooling systems reach peak load very occasionally and then only for a short period of time. Most of the time, the equipment is operating at loads much below peak. Equipment either improves in efficiency at lower load—a characteristic of many boilers—or it falls—a characteristic of many refrigeration units. When choosing refrigeration equipment, it can be very worthwhile to consider the part-load performance. It is in the part-load performance evaluation that hour-by-hour computer simulations become a really necessary tool.
>
> **Load versus capacity**. Note that we have been talking about "loads," but when you look in manufacturers' data sheets, they talk about "plant capacity." "Loads" and "capacity" are the same issue, but **loads** are the calculated building requirements, while **capacity** is the plant equipment's ability to handle the load. When purchasing packaged plant equipment, the plant capacity often does not exactly match the calculated building loads. One of the challenges for the designer is choosing the most suitable package, even though it does not exactly match the calculated building loads. This issue is illustrated in the following section on rooftop units.

## 6.6 Rooftop Units

A typical rooftop system is diagrammed in *Figure 6.8*. The return air is drawn up into the base of the unit and the supply air is blown vertically down from the bottom of the unit into the space below. As an alternative, the ducts can project from the end of the unit to run across the roof before entering the building.

The major advantages of these units are

> *No working parts in the occupied space*—so maintenance can be carried out without disrupting activities within the building and maintenance can be carried out without access to the building when the building is closed.
> *No space is built for the unit*—which saves construction costs.
> *No delay for detailed manufacturer design work*—because the unit is pre-designed.
> *No wide access during construction*—because the unit is outside the building envelope, the contractor does not have to keep an access available for the unit to be moved in during construction.

There are, of course, disadvantages.

> *Critical units must be maintained regardless of the weather conditions*—That means that maintenance could be required in heavy rain, snow, or high

Single Zone Air Handlers and Unitary Equipment    83

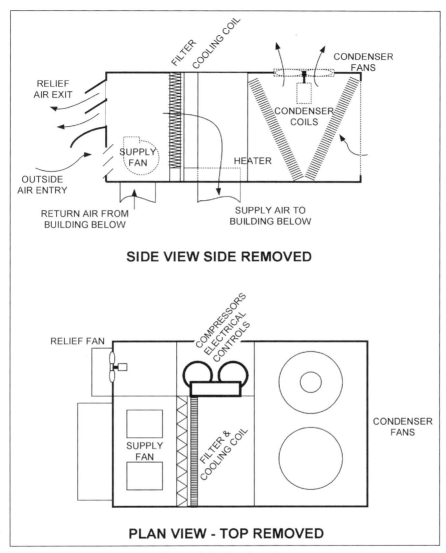

**Figure 6.8**  Rooftop Unit

winds. This potential problem can be managed by having a maintenance access space located along one side of the unit.

*Choice of performance is limited to the available set of components*—This is often not enough of a problem to make the unit unacceptable, and can frequently be overcome by using a split unit, which we will be discussing in the next section.

Choosing a rooftop unit is fairly straightforward. One needs to know both inside and outside design-temperatures, required airflow, in cfm, mixed-air temperature, and the required sensible and latent cooling-loads.

The mixed-air temperature can be calculated based on the return-air temperature, the-outside air temperature and the required proportion of outside air. Referring back to the example, shown in *Figure 6.7* above, the room temperature, which we will consider to be return temperature, was 75°F, and the outside ambient temperature was 95°F. If 20% outside air is required, then the mixed temperature can be estimated by proportion

95°F · 0.2 + 75°F · 0.8 = 79°F

The calculation of airflow is covered in detail in ASHRAE Course *Fundamentals for Air System Design*[1].

It is important that the airflow is correctly calculated and that the unit is setup and balanced to provide the correct airflow. With direct expansion refrigeration circuits, too little airflow over the evaporator can cause problems:

Imagine that the airflow is much slower than design. The slow speed past the coil will allow the air to cool further, and if the coil is below freezing, for ice to start to form. The slow flow will also reduce the heat being absorbed into the evaporator, so the compressor's suction will be drawing with little refrigerant vapor coming in. As a result, the pressure in the evaporator will fall, causing the evaporator temperature to fall, which again, will tend to cause freezing. Once ice formation starts, the ice starts to block the flow, causing even slower airflow until the coil is encased in ice. Ice formation on the evaporator can also be caused by too little refrigerant in the system – a common result of a slow refrigerant leak.

As noted in the previous discussion of loads versus capacity, air-handling units come in discrete sizes, so a perfect match of unit and calculated loads does not happen. From the example in *Figure 6.7*, for our loads of 200,000 Btu/h sensible load and 20,000 Btu/h latent load, let us assume the closest unit has a performance of 240,000 Btu/h sensible and 60,000 Btu/h latent capacity. This looks excessively oversized, but two factors have to be considered: First the unit's sensible capacity does not take into account the heat from the supply fans in the unit. Suppose the fan load was 6 kW (3,412 Btu/hr = 1 kW) then the fan-heat added to the cool air would be:

6 · 3412 Btu/h = 20,472 Btu/h

The effective sensible heat capacity of the unit is thus:

240,000 Btu/h − 20,472 Btu/h = 219,528 Btu/h

This is a very close match to the required capacity.

The 60,000 Btu/h moisture removal, when compared to the required 20,000 Btu/h, is a common issue in dry climates. The coil removes more moisture than required. There are two results. First, more energy is used than required to maintain the design conditions. Second, the real conditions will be drier than the design condition.

The converse problem, of too little moisture removal, occurs in hot moist climates, particularly where higher proportions of outside air are required. In this case, and others, it may not be possible to find a package rooftop-unit for the duty and it may be advantageous, or necessary, to take special measures to remove moisture. Some of these are discussed in Chapter 13.

Heating choices are generally less of an issue, but the designer still has to be aware of potential problems. As noted earlier, electrical heaters are normally available with stepped capacity, but gas heaters are often on-off or high-low-off. If the unit runs continuously as the gas-heater cycle, the air supply will fluctuate in temperature and sometimes blow warm, and sometimes blow cold. Take care to ensure that the occupants do not have an intermittent cold draft blowing on them.

Having considered the single-zone air handler, with particular emphasis on the rooftop unit, let us now consider another popular single-zone system, the split system.

## 6.7 Split Systems

In the rooftop unit, all the plant was in a single housing and was purchased as a manufacturer's pre-design. In general, the package rooftop-units are designed for popular duties, and to be as light and compact as possible, since they have to be lifted onto the roof. In the split system, the compressor/condenser part of the refrigeration system is chosen separately from the rest of the system and connected by the refrigerant lines to the air system, which includes the evaporator. The pipes, even with their insulation, are only inches in diameter, compared to ducts that are, typically, feet in diameter. The separation of the two parts of the refrigeration system to produce the split system is diagrammed in *Figure 6.9*. The system can range in size from the small residential systems where the inside coil is mounted on the furnace air outlet to substantial commercial units serving a building.

The split system allows the designer a much greater choice of performance. For example, designing a unit for operation in an ice rink requires a low space temperature, hence a non-package situation. This requirement is well suited to the flexibility of the split system.

The other main advantage of the split system is that it allows the air handling part of the unit to be indoors, where it is easier to maintain and does not need to be weatherproofed. The noise of the compressor is outside and can be located at some distance from the air-handling unit. For example, in a three-storey building, all the condensers can be mounted on the roof, while the air handlers are on the floor they serve. This allows the ducting to be run horizontally on each floor and only requires a small vertical duct for the refrigerant lines from the three units to the roof.

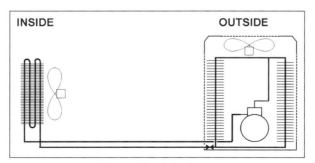

**Figure 6.9** Split System

## The Next Step

We have considered single zone air-conditioning systems in this chapter. We focused on rooftop and split systems. We considered the components they contain, how the components operate and some of the limitations of off-the-shelf equipment. Finally we looked at a simple choice of rooftop and spilt system and the resulting space conditions.

In the next chapter we will look at how these single zone systems can be modified to produce multi-zone systems.

## Summary

In this chapter, we discussed issues of system choice and provided a general description of system control issues. We will return to controls in more depth in Chapter 11.

### 6.2 Examples of Buildings with Single zone Package Air-Conditioning Units

For heating and cooling, a packaged unit may require: just an electrical source of power, or a gas or hot water supply, and/or a source of chilled water. The basic operation of the unit stays the same; it is just the source of heating and cooling energy that may change.

### 6.3 Air Handling Unit Components

The overall functions of the air-handler are to draw in outside air and return air, mix them, condition the mixed air, blow the conditioned air into the space and exhaust any excess air to outside. Components of the unit can include: inlet louver screen, the parallel blade damper, opposed blade damper, the relief air damper, actuator, the mixed temperature sensor, filter heating coil, cooling coil, humidifier, fan, return fan. The concept of control logic was introduced as a method to summarize the operation of the components of the system.

### 6.4 Refrigeration Equipment

The vapor compression refrigeration cycle is generally the basis of mechanical refrigeration. The vapor compression refrigeration system comprises four components: compressor, condenser, expansion valve, and evaporator. This system can be used directly, to provide cooling to, typically, a local coil. To provide cooling for several coils at greater distances, refrigeration machines are used to chill water that is then pumped around buildings to provide cooling in air-conditioning systems. The heat removed from the water is expelled into the atmosphere through a hot, air-cooled coil, or by evaporating water in a cooling tower.

The components are matched to work together with a specific charge of refrigerant. If you operate the system with too little refrigerant or too little air or water flow over the evaporator or condenser, problems can arise.

While cooling is achieved by pumping heat from a cooled space and rejecting heat to outside, you can reverse the process, in a mild climate, with a heat pump, to obtain heat from ambient air. Similarly, the ground can be used as a source of heat or a sink for waste heat, by using a ground source heat pump.

### 6.5 System Performance Requirements

Before choosing a system, you need an understanding of the types of loads you want the system to manage. Summer cooling loads will be the main determinant of the choice of unit. These summer factors are used to determine the summer load: outside design temperature; outside design humidity; inside design temperature and humidity; inside generation of heat and moisture; ventilation requirements. Once you have determined summer loads, additional decision factors for unit choice are the initial cost to purchase and install, versus the ongoing cost of operation and maintenance; and load versus capacity.

### 6.6 Rooftop Units

In a typical rooftop unit, the return air is drawn up into the base of the unit and the supply air is blown vertically down from the bottom of the unit into the space below. As an alternative, the ducts can come out of the end of the unit to run across the roof before entering the building. Advantages and disadvantages of rooftop units were discussed.

Choice factors to choose a rooftop unit: inside and outside design temperatures, required airflow in cfm, mixed air temperature, and the required sensible and latent cooling loads.

It is important that the airflow is correctly calculated and that the unit is setup and balanced to provide the correct airflow. With direct expansion refrigeration circuits, too little airflow over the evaporator can cause icing problems.

Units come in discrete sizes so a perfect match of unit and calculated loads does not happen. As a result, the design conditions may be jeopardized, and or extra energy costs may arise.

### 6.7 Split Systems

In the split system, the compressor/condenser part of the refrigeration system separate from the evaporator coil and connected by the refrigerant lines to the air system, which includes the evaporator.

Advantages of the split system: It allows the designer a much greater choice of performance; it allows the air handling part of the unit to be indoors, where it is easier to maintain and does not need to be weatherproofed. The noise of the compressor is outside and can be located at some distance from the air-handling unit.

## Bibliography

1. ASHRAE Fundamentals of Air System Design
2. ASHRAE Fundamentals of Refrigeration
3. ASHRAE Fundamentals of Heating and Cooling Loads

Chapter 7

# Multiple Zone Air Systems

## Contents of Chapter 7

Study Objectives of Chapter 7
7.1  Introduction
7.2  Single-Duct, Zoned Reheat, Constant-Volume Systems
7.3  Single-Duct, Variable-Air-Volume Systems
7.4  By-pass Box Systems
7.5  Constant Volume Dual-Duct, All-Air Systems
7.6  Multizone Systems
7.7  Three-Deck Multizone Systems
7.8  Dual-Duct, Variable-Air-Volume Systems
7.9  Dual Path Outside Air Systems
The Next Step
Summary

## Study Objectives of Chapter 7

Chapter 7 shows the most common ways that a single-supply air system can be adapted to provide all-air air conditioning to many zones with differing loads. After studying the chapter, you should be able to:

Identify, describe and diagrammatically sketch the most common all-air air-conditioning systems.
Understand the relative efficiency or inefficiency of each type of multiple zone air system.
Explain why systems that serve many zones, and that have a variable-supply air volume, are more energy-efficient than those with constant-supply volumes.

## 7.1 Introduction

In the last chapter, we considered two types of single zone direct expansion systems: the packaged rooftop system and the split system. The direct-expansion-refrigeration rooftop unit contained all the necessary components to condition a single air supply for air-conditioning purposes.

These same components can be manufactured in a wide range of type and size. As an alternative to a rooftop unit, they can be installed indoors, in a mechanical room, with the different components connected by sheet-metal ducting.

Both the packaged rooftop unit and the inside, single-zone unit produce the same output: a supply of treated air at a particular temperature.

The heating or cooling effect of this treated airflow, when it enters a zone, is dependent upon two factors:

The flow rate, (measured in cubic feet per minute, cfm).
The temperature difference between the supply air and the zone temperature, (measured in degrees Fahrenheit, °F).

When the unit is supplying one space, or zone, the temperature in the zone can be controlled by

Changing the air volume flow rate to the space.
Changing the supply air temperature.
Changing both air volume flow and supply air temperature.

In many buildings, the unit must serve several zones, and each zone has its own varying load. To maintain temperature control, each zone has an individual thermostat that controls the volume and/or temperature of the air coming into the zone.

Air-conditioning systems that use just air for air conditioning are called "all-air systems".

These all-air systems have a number of advantages:

**Centrally located equipment**—operation and maintenance can be consolidated in unoccupied areas, which facilitates containment of noise.
**Least infringement on conditioned floor space**—conditioned area is free of drains, electrical equipment, power wiring and filters (in most systems).
**Greatest potential for the use of an economizer cycle**—as discussed in Chapter 2, this can reduce the mechanical refrigeration requirements by using outside air for cooling, and therefore reduce overall system operating costs.
**Zoning flexibility and choice**—simultaneous availability of heating or cooling during seasonal fluctuations, like spring and fall. The system is adaptable to automatic seasonal changeover.
**Full design freedom**—allows for optimum air distribution for air motion and draft control.
**Generally good humidity control**—for both humidification and dehumidification.

All-air systems generally have the following disadvantages:

**Increased space requirements**—significant additional duct space requirements for duct risers and ceiling distribution ducts.
**Construction dust**—due to problems with construction-dust, all-air systems are generally available for heating later in the construction schedule than systems that use water to convey heat.
**Closer coordination required**—all-air systems call for close cooperation between architectural, mechanical and structural designers.

In addition to these general disadvantages, constant-volume-reheat systems are particularly high energy consumers because they first cool the air, and then reheat it. Because the reheat coils are sometimes hot water coils, an additional potential disadvantage is a problem with leaking hot-water coils. We will discuss these systems in more detail in the next section.

To make these all-air systems work for many zones requires some form of zone control. In this chapter we will consider how zone control can be achieved with all-air air-conditioning systems.

The simplest, and one that we will start with, is the **constant-volume-reheat** system.

## 7.2 Single-Duct, Zoned Reheat, Constant Volume Systems

The **reheat system** is a modification of the single-zone system. The reheat system permits zone control by reheating the cool airflow to the temperature required for a particular zone. *Figure 7.1* shows a reheat system, with ceiling supply diffusers in the space.

A constant volume of conditioned air is supplied from a central unit at a normally, fixed temperature, (typically 55°F). This fixed temperature is designed to offset the maximum cooling load in all zones of the space. If the actual cooling load is less than peak, then the reheat coil provides heat equal to the difference between the peak and actual loads. When heating is required, the heater heats the air above zone temperature to provide heating.

The reheat coil is located close to the zone and it is controlled by the zone thermostat. Reheat coils are usually hot water or electric coils. As noted above, if the reheat coils are hot water, then there can be a problem with leakage.

A reheat system is often used in hospitals, in laboratories, or other spaces where wide load-variations are expected.

When primary air passes quickly over a vent, it draws some room air into the vent. This process is called **induction**. There are two variations on the reheat system that both use **induced** room air: the **Induction Reheat Unit**, shown in *Figure 7.2*; and the **Low-Temperature Reheat Unit with Induced Air** shown in *Figure 7.3*.

The **Induction Reheat Unit** shown in *Figure 7.2* shows the primary supply of air, blown into the unit and directed through the induction

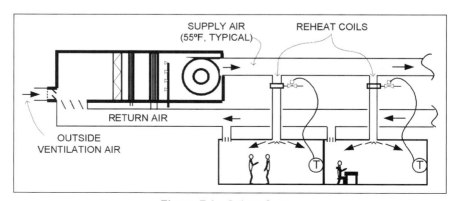

**Figure 7.1** Reheat System

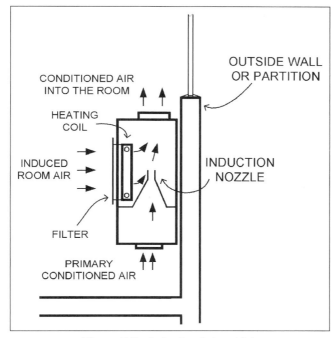

**Figure 7.2** Induction Reheat Unit

nozzle. The reduced aperture of the nozzle forces the air to speed up and move quickly to the unit exit, into the room. As the primary air passes quickly past the reheat coil, it draws, or induces, air from the room into the unit. The room air passes across the reheat coil and mixes with the primary air.

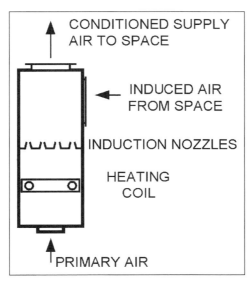

**Figure 7.3** Low-Temperature Reheat Unit with Induced Air

Units like this are often mounted beneath windows, where they offset any downdraft in cold weather. In addition, even when the air supply is turned "off," hot water in the coil will still provide some heating.

The second type of induction reheat system, the **Low-Temperature Reheat Unit with Induced Air**, shown in *Figure 7.3*, is used where very cold supply air is provided. In some systems, the supply air can be as cold as 40°F. This could create intolerable drafts and serious condensation on the supply outlets. In this system, the primary air is preheated when necessary, but room air is always induced to mix with the primary air to ensure that the flow into the space is not excessively cold.

There are two primary advantages to this system:

- Duct sizing: When the system is designed to use 40°F supply air, ducts can be sized for half the air volume, compared to the ducts required for a 55°F supply-air temperature. This results in a lower installation cost, and a smaller requirement for duct space.
- The small volume of supply air may be exhausted from the room rather than returned to the main cooling system, possibly eliminating the need for return ductwork.

Overall, reheat systems are simple, and **initial costs**, the costs of design and construction, are reasonable. Reheat systems provide good humidity control, good temperature control, good air circulation, and good air quality.

The problem with all reheat systems is their energy inefficiency, so they are expensive systems to run. Generally, when the load is less than the peak cooling load, the cooling effect and the reheat are working against each other to neutralize their contributions. This means, in a no-load situation, the refrigeration is going at full blast and the reheat is just matching the cooling effect. There are two energy drains for no load! This is not quite as severe as it sounds because the no-load condition is the worst-case scenario, and it only occurs for a relatively small amount of the time.

Overall, though, reheat is energy expensive. As a result, these systems have fallen out of favor in recent times.

## 7.3 Single-Duct, Variable Air Volume Systems

Buildings that are located in continuously warm climates, and interior spaces in any climate, require no heating, only cooling. For cooling-only situations, it would be ideal to supply only as much cooling and ventilation as the zone actually requires at the particular moment. A system that comes close to the ideal is the **variable-air-volume** system "**VAV**".

The variable air volume system is designed with a volume control damper, controlled by the zone thermostat, in each zone. This damper acts as a throttle to allow more or less cool air into the zone. The VAV system adjusts for varying cooling loads in different zones by individually throttling the supply air volume to each zone. Regardless of the variations in the cooling load, a minimum flow of ventilation air is always provided and care must be taken to ensure that the required volume of ventilation air is provided.

In a VAV system, as the zone becomes cooler, the cooling load decreases and the cool airflow to the zone decreases. Eventually it reaches the minimum value necessary for adequate ventilation and air supply, *Figure 7.4*. When this

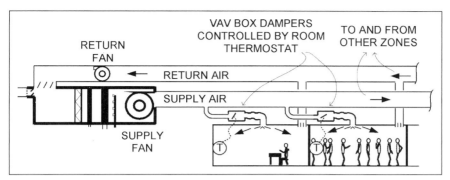

**Figure 7.4** Variable Air Volume System

minimum airflow is reached, if the zone is still too cool, heating is provided by a thermostatically controlled reheat coil or a baseboard heater.

This means there may be some energy wasted in the VAV system, due to heating and cooling at the same time. However, this energy waste is far less than in the terminal reheat system, since the cooling ventilation air is reduced to a minimum before the heating starts.

The total supply-airflow rate in a VAV system will vary as the zone dampers adjust the flow to each zone. Therefore, the supply fan must be capable of varying its flow rate. The variation in flow rate must be achieved without allowing the duct pressure to rise excessively or to drop below the pressure required by the VAV boxes for their proper operation. This pressure control is often achieved by using a pressure sensor in the duct to adjust a fan-speed control unit. Similarly, the return fan is controlled to meet the varying supply-air volume.

There are other methods that are discussed in the ASHRAE Course, Fundamentals of Air System Design.

In systems where the fan speed is reduced to reduce the volume flow, the fan power drops substantially as the flow reduces. This reduction in fan power is a major contribution to the economy of the VAV system.

VAV systems may have variable volume return air fans that are controlled by pressure in the building or are controlled to track the supply-fan volume flow.

In small systems, the variable-volume supply may be achieved by using a relief damper, called a "**bypass**," at the air-handling unit. The bypass allows air from the supply duct through a control damper into the return duct, as shown in *Figure 7.5*.

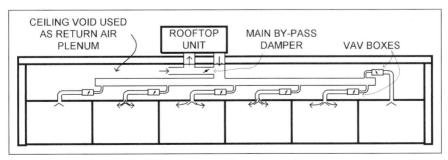

**Figure 7.5** Variable Air Volume System With Bypass

As the zones reduce their air requirements, the bypass damper opens to maintain constant flow through the supply fan. This arrangement allows for the constant volume required by the refrigeration circuit. For smaller systems, this method can provide very effective zone control without creating problems that may occur when the airflow is varied across the direct expansion refrigeration coil. Unfortunately, this system keeps the fan working at near full load.

### VAV Advantages

Advantages of the variable volume system are the low initial costs and low operating costs. Initial costs are low because the system only requires single runs of duct and a simple control at the end of the duct. Operating costs are low because the volume of air, and therefore the refrigeration and fan power, closely follow the actual load of the building. There is little of the cool-and-reheat inefficiency of the reheat system.

### VAV Problems

There are potential problem areas with variable air volume systems. These include: poor air circulation in the conditioned space at lower flows; dumping of cold air into an occupied zone at low flows; and inadequate fresh air supplied to the zone. Improved diffusers have made it possible for the designer to avoid dumping and poor room circulation. However, the problem of inadequate outside air for ventilation needs additional care when the system is being designed.

For example, as we saw in the last chapter, in a constant volume system where all the zones require 20% outside air, setting the outside air to 20% on the main unit ensures that each zone receives 20% outside air. In the VAV system, one cannot set the outside air proportion. As the zone flows are reduced due to low thermal load, the proportion of outside-air-for-ventilation needs to increase. As a result, the outside-air volume must be maintained at all volume flows. This can be achieved in a number of ways, but the process requires a sophisticated, and potentially more expensive control system that is not required in constant volume systems.

## 7.4 By-pass Box Systems

Where the main supply unit must handle a constant volume of air, **by-pass boxes** can provide a variable volume of air to the zones served. The bypass boxes can be used on each zone, or as you saw in Figure 7.4, a single central by-pass can be used with variable volume boxes serving each zone.

*Figure 7.6* shows the use of the by-pass box on each zone. A thermostat in each zone controls the damper in the by-pass box serving the zone. The flow of air to each box is essentially constant. The bypass box, shown on the left, is set for full flow to the zone. The box in the center is passing some air to the zone and bypassing the balance. The zone on the right is unoccupied, and the box is set to bypass the full flow. The zone thermostat controls how much of the air is directed into the zone and how much is by-passed into the return-air system. In many buildings, the return can be via the space above the dropped ceiling, the **ceiling plenum**, and then, via a duct, back to the return of the air-handling unit.

With the by-pass system, it is important to keep the ceiling plenum at a negative pressure, so that the excess cooling air does not leak into the zone.

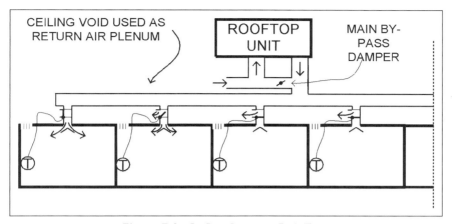

**Figure 7.6** By-Pass Boxes on Each Zone

The danger of keeping the ceiling at negative pressure, though, is that this can cause infiltration of outside air through the walls and roof joints, resulting in moisture and load challenges.

## 7.5 Constant Volume Dual-Duct, All-Air Systems

**A dual-duct system** employs a different approach for establishing zone control. In a dual-duct system, cooling and heating coils are placed in separate ducts, and the hot and cold air flow streams are mixed, as needed, for temperature control within each zone.

In this system, the air from the supply fan is split into two parallel ducts, downstream of the fan. One duct is for heating and the other for cooling. A layout of three zones of a dual-duct system is shown in *Figure 7.7*.

The duct with the heating coil is known as the hot deck, and the duct with the cooling coil is the cold deck. These constant volume dual-duct systems usually use a single, constant-volume supply fan to supply the two ducts.

The dual-duct system can also be drawn diagrammatically as shown in *Figure 7.8*. Satisfy yourself that the two figures show the same system, although they look very different.

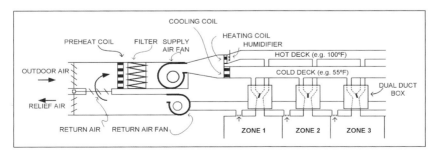

**Figure 7.7** Dual-Duct System, Double Line Diagram

96   Fundamentals of HVAC

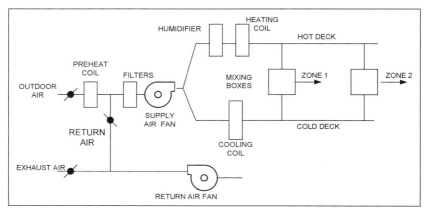

**Figure 7.8**  Dual-Duct System, Single Line Diagram

Dual-duct systems achieve the zoned temperature control by mixing the hot and cold air streams in a dual-duct box while maintaining a constant airflow. As in the reheat system described earlier, the heating and cooling effects are fighting against each other when the load is less than peak load. The combined energy use leads to energy inefficiency, which is the biggest disadvantage of dual-duct systems. The energy inefficiency may be reduced by these methods:

Minimizing the temperature of the hot deck using control logic based on zone loads or outside temperature
Raising the cold deck temperature when temperature and humidity conditions make it practical
Using variable volume dual-duct mixing boxes

The system also has a high first cost, since it requires two supply ducts. These two ducts need additional space above the ceiling for the second supply duct and connections.

Dual-duct systems were popular in the 1960s and 1970s and many are installed in hospitals, museums, universities, and laboratories. Due to the relatively high installation and operating costs, dual-duct systems have fallen out of favor except in hospitals and laboratories, where their ability to serve highly variable sensible-heat loads at constant airflow make them attractive. Another advantage of dual-duct systems is that there are no reheat coils near the zones, so the problems of leaking hot water coils is avoided.

The dual-duct system delivers a constant volume of air, with varying percentages of hot and cold air, as shown in *Figure 7.9*.

In Figure 7.9, there are plots of percentage flow from the hot and cold air streams as a function of room temperature. The sum of the hot and cold airstream percentages always adds up to 100%. For the room temperature setpoint range, also known as the throttling range, of 70°F to 72°F, the thermostat will control the hot-air flow linearly, from 100% at 70°F to 0% at 72°F. Outside the throttling-temperature range, the flow is either all hot air or all cold air.

In *Figure 7.10*, there is a different view of the same process over the throttling range.

There are two plots. One plot, the solid line, shows how the delivered air temperature will vary as the thermostat controls the percentage mixture of hot

Multiple Zone Air Systems    97

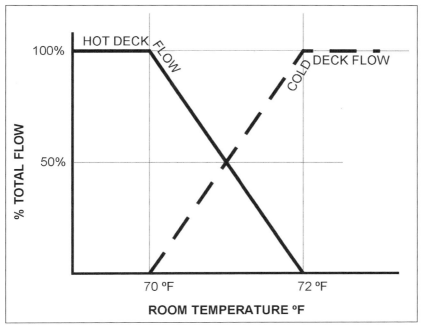

**Figure 7.9** Air Flow in a Dual-Duct System

and cold streams. The delivered air-temperature scale is on the right-hand side of the graph, and the room-temperature scale is on the horizontal axis.

At a room temperature of 70°F and below, with 100% hot air, the delivery temperature is at 110°F. At a room temperature of 72°F and above, with 100%

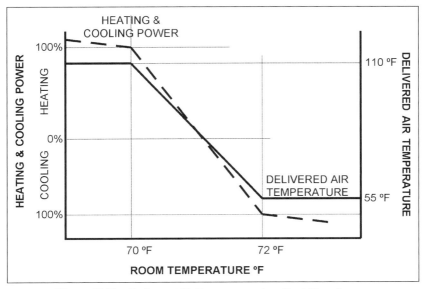

**Figure 7.10**   Delivered Air Temperature in a Dual-Duct System

cold air, the delivery temperature is 55°F. At room temperatures between 70° and 72°F, the delivery temperature varies linearly with the room temperature.

The second plot in *Figure 7.10*, the dashed line, is that of the net cooling or heating power delivered to the zone to meet the load. The scale for the power variable is on the vertical axis, on the left-hand side of the graph. Zero power, (or no net delivered heating or cooling) is at mid-height on the vertical axis. Above the mid-height, there is net heating and below mid-height, there is net cooling.

It is important to observe that, because this is a constant volume system, *zero power* does **not** mean *zero energy use*. Zero power corresponds to an equal amount of heating and cooling, so that the heating and cooling effects cancel each other out, and give a neutral temperature effect on the zone.

As shown in *Figure 7.9*, below a room temperature of 70°F, the flow is 100% heating at 110°F; and above a room temperature of 72°F, the flow is 100% cooling at 55°F. Between 70° and 72°F, the flow is a linear mixture of hot and cold air.

## 7.6 Multizone Systems

The multizone system is thermodynamically the same as the dual-duct system. They both involve mixing varying proportions of a hot-air stream with a cold-air stream to obtain the required supply temperature for that zone. In the dual-duct system, the mixing occurs close to the zone, in the dual-duct box. In the multizone system, as shown in *Figure 7.11*, the mixing occurs at the main air-handling unit.

The basic multizone system has the fan blowing the mixed air over a heating coil and a cooling coil in parallel configuration. As you know, in the dual-duct system, the resulting hot and cold air is ducted through the building to dual-duct mixing boxes. In contrast, in the multizone system, the heating and cooling airflows are mixed in the air-handling unit at the coils using pairs of dampers.

The hot deck coil is arranged above the cold deck coil and they are sectioned off into zones; just two sections are shown in the figure. Each section has a two-section damper that opens to the cold deck as it closes to the hot deck. Each damper pair is driven by an actuator pushing the crank at the end of the damper shaft. The mixed air from each section is then ducted to a zone.

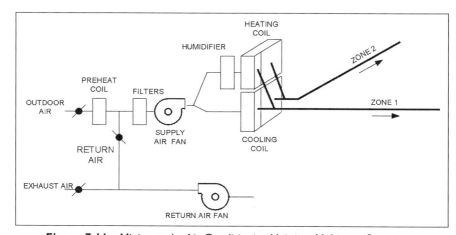

**Figure 7.11** Mixing at the Air Conditioning Unit in a Multizone System

As in the dual-duct system, a certain amount of energy inefficiency occurs because the air is being both heated and cooled at the same time.

## 7.7 Three-deck Multizone Systems

The three-deck multizone system is a possible solution to overcome the energy inefficiency of the overlapping use of heating and cooling in a traditional multizone system.

The three-deck system is similar to the dual-duct and multizone systems, except that there is an additional (third) air stream that is neither heated nor cooled. Hot and cold air are never mixed in the three-deck system. Instead, thermal zones that require cooling receive a mixture of cold and neutral air, and thermal zones that require heating receive a mixture of hot and neutral air. The air flow control is shown in *Figure 7.12*. Thus, the three-deck system avoids the energy waste due to the mixing of hot and cold air streams.

The neutral air in the three-deck system is neither heated nor cooled and its temperature will change with the season. In summer, the neutral air will be warmer than the cold deck air. Consequently, the neutral air will take the place of the hot-deck air, eliminating the need for the heating coil in summer. In winter, the neutral air will be cooler than the hot deck, thus replacing the cold deck and the need for activating the cooling coil in winter. The net annual result is that there is no penalty for having heating and cooling coils operating simultaneously.

## 7.8 Dual-Duct, Variable Air Volume Systems

The dual-duct, variable air volume (VAV) system provides the thermal efficiency of the VAV system while generally maintaining higher air flows, and thus better circulation of air in the room, when heating is required. The difference is that the air is not drawn into the building by a constant volume fan, as it is in the usual dual-duct system, but it is split into two air streams that flow through two variable-volume fans. One air stream passes through

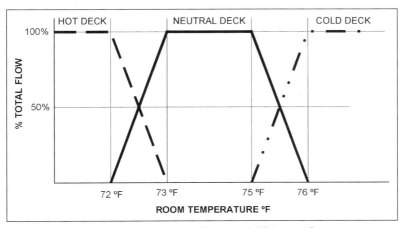

**Figure 7.12** Air Flow for Three-deck, Multizone System

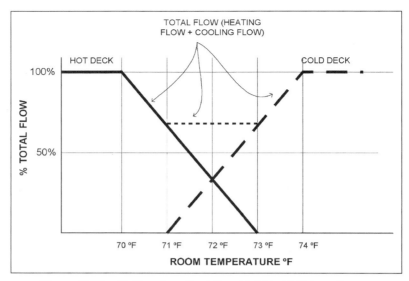

**Figure 7.13**   Air Flow for a Dual-Duct, Variable Air Volume System

a heating coil and one through a cooling coil. The two air streams are then ducted throughout the building.

The mixing of these two air streams is carried out in a mixing box serving each thermal zone. These mixing boxes can vary both the proportions of hot and cold air, and also the total flow rate of air to the zone. This is in contrast to the more conventional dual-duct system where the airflow delivered by the mixing box is constant.

The variation of flow in the dual-duct, variable-air-volume system is shown in *Figure 7.13*. This diagram indicates equal volume flows for both heating air and cooling air. Depending on the climate and resulting loads, the heating flow many be 50% less than the cooling airflow, but the control logic is the same. At maximum cooling load, the box provides sufficient cold air to meet the load. As the cooling load decreases, the volume of cold air is decreased, without addition of hot air to change the temperature. When the cooling load reaches the point where the cold airflow equals the minimum allowable flow, the cold flow continues to decrease, but the hot air is added to maintain sufficient total flow. As the heating load increases, the total flow remains constant while its temperature is increased above room temperature by increasing the proportion of air from the hot deck. When the cold deck flow reaches zero, the temperature of the delivered air will be the hot deck temperature. As the heating load increases further, the requirement for more heat is satisfied by increasing the volume flow-rate of hot air.

## 7.9 Dual Path Outside Air Systems

Throughout this text, our examples have shown the outside ventilation air being mixed with return air before being processed and supplied to the building. This mixing method works well in cooler, dryer climates. This does not work as well in warm/hot, humid climates. The reason is very simple: the main cooling coil cannot remove enough moisture without overcooling the whole air stream. What is required is high moisture removal without full cooling.

An effective way around this problem is to use a dual path system. The outside air comes in through a separate, dedicated cooling coil before mixing with the return air. This dedicated outdoor air coil has two functions.

> *Dehumidification*: The system is designed and operated to dehumidify the outside air to a little below the required space-moisture content.
> *Cooling*: The system cools the outside air to about the same temperature as the main coil, when the main coil is at maximum cooling.

When the system is in operation, the fully cooled outside air, say 20%, mixes with 80% return air before it reaches the main cooling coil. The mixture is equivalent to the full airflow, substantially dehumidified and 20% cooled. The main cooling coil now provides the required extra cooling that the system needs, and a modest, achievable, requirement for dehumidification.

The challenge of providing adequate dehumidification at an acceptable cost is an ongoing challenge in moist climates. The dual path method described above is one of the many ways available to tackle the challenge of removing moisture without overcooling.

## The Next Step

This chapter has been all about all-air systems that serve many zones. In many cases systems with separate water heating and or cooling can be very effective. For instance, in a very cold climate, it is often more comfortable to provide a perimeter hot water heating system and use the air system for cooling, ventilation air supply, and fine temperature control. This also allows the air system to be turned off when the building is unoccupied, even though the heating system must remain on to prevent over-cooling or freezing.

In the next chapter, Chapter 8 we will consider water systems and how they coordinate with air systems we have discussed in this chapter and the previous one.

## Summary

This chapter has introduced the various ways zoning can be achieved with all-air air-conditioning systems. They are all based on individually varying the air flow and/or temperature supplied to each zone.

### 7.2 The Reheat System

Reheat is the simplest system, known for both its reliability and the down side, its high energy wastage. Two induction variations were introduced: one that also provides some night time heating; and the other that accommodates very low supply-air temperatures.

102    Fundamentals of HVAC

## 7.3  Variable Air Volume, VAV System

More energy efficient than reheat, VAV is a very flexible system with many virtues. When there is a low load, however, it does offer challenges for maintaining adequate ventilation air and good room air distribution.

## 7.4  The Bypass System

A variation on the VAV system, the bypass system, is suitable for providing good control in smaller systems, and for constant flow over a direct-expansion cooling coil. Designers must be cautious to ensure that bypassed air goes straight back to the air conditioning unit, but it is generally a simple system to design.

## 7.5  The Dual-Duct System

The system provides full airflow when the system is on, but, like the reheat system, suffers from the energy penalty of simultaneous heating and cooling. A very attractive feature of the dual-duct system is that there are no reheat coils near the zones, so the problems of leaking hot water coils is avoided.

## 7.6  The Multizone System

A system thermodynamically similar to the dual-duct system, the multizone system features a different layout. The multizone system is not as energy efficient as the VAV system, and requires a separate duct to each zone. However, the multizone system has the advantage of requiring no maintenance outside the mechanical room, except for the zone temperature-sensors and associated cable.

## 7.7  Three-deck Multizone System

The more modern introduction of the third, neutral duct to the multizone system, avoids the conflict of concurrent heating and cooling.

## 7.8  Dual-Duct, Variable Air Volume System

A modification of the dual-duct system, this system uses variable volume dual-duct boxes to provide the thermal efficiency of the VAV system, while maintaining higher air flows, and thus better room air circulation when heating is required.

## 7.9  Dual Path Outside Air System

This system could be used to reduce the problem with excess moisture in the air that arises in warm/hot, humid climates.

Chapter 8

# Hydronic Systems

## Contents of Chapter 8

Study Objectives of Chapter 8
8.1  Introduction
8.2  Natural Convection and Low Temperature Radiation Heating Systems
8.3  Panel Heating and Cooling
8.4  Fan Coils
8.5  Two Pipe Induction Systems
8.6  Water Source Heat Pumps
The Next Step
Summary
Bibliography

## Study Objectives of Chapter 8

Chapter 8 introduces hydronic systems, which are also known as water systems. **Hydronic systems**, in this text, are systems that use water or steam as the heat transfer medium. In some places, the term "hydronic" has become associated with just radiant floor heating systems, which is a rather narrower definition than we are using in this text. We will discuss radiant floor heating systems in 8.3: Panel Heating and Cooling.

Hydronic systems have their own characteristics, benefits and challenges. After studying the chapter, you should be able to:

Describe five types of hydronic systems
Explain the main benefits of hydronic systems
Discuss some of the challenges of hydronic systems
Explain the operation and benefits of a water-source heat pump system

## 8.1 Introduction

In the previous two chapters, we discussed single zone and multiple-zone all-air air-conditioning systems. In Chapter 7, Section 7.2, we mentioned that water coils could be used in the main air-handling unit and for the reheat coils in the reheat and VAV systems. In this chapter we are going to consider systems where water-heated and/or water-cooled equipment provide most of the heating and/or cooling.

In some buildings, these systems will use low-pressure steam instead of hot water for heating. The performance is generally similar to hot water systems, with higher outputs due to the higher temperature of the steam. However, control in these steam systems is generally inferior, due to the fixed temperature of steam. For steam systems and boilers see Chapters 10 and 27 respectively of *ASHRAE 2000 Systems and Equipment Handbook*. The properties of steam, the theory of two-phase flow and steam pipe sizing, are covered in Chapters 6, 4, and 35 of *ASHRAE 2001 Fundamentals Handbook*.

Throughout the rest of this chapter, we will assume that hot water is being used as the heating medium.

Because of their ability to produce high output on an 'as-needed basis,' hydronic systems are most commonly used where high and variable sensible heating and/or cooling loads occur. These are typically

- Perimeter zones, with high solar heat gains or
- Perimeter areas in cooler to cold climates where there are substantial perimeter heat losses.

The entrance lobby of a building in a cold climate is an example of an ideal use for these systems. They are frequently used in office buildings, hospitals, hotels, schools, apartment buildings and research laboratories in conjunction with ventilation and cooling air systems.

*Hydronic systems advantages*:

Noise reduction—Virtually silent operation
Economy, due to limited operational costs—Large amounts of heat from small local equipment
Economy due to limited first costs—Pipes are small compared to ducts for the same heat transfer around a building
Energy efficiency—Low energy consumption at low load

*Hydronic systems disadvantages*:

Ventilation—Provision of outside air for ventilation is either absent or poor
System failure—Danger from freezing and from leaks
Humidity—Control is either absent or generally poor

We will start our discussion with simple heating systems that operate by allowing heat to escape from a hot surface by natural convection and low temperature radiation.

## 8.2 Natural Convection and Low Temperature Radiation Heating Systems

The very simplest water heating systems consist of pipes with hot water flowing through them. The output from a bare pipe is generally too low to be effective, so an extended surface is used to dissipate more heat. There is a

Hydronic Systems 105

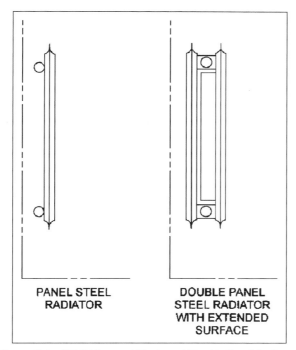

**Figure 8.1** Wall-Mounted Single and Double Panel Radiators

vast array of heat emitters. A small selection of types is shown in *Figures 8.1* and *8.2*. Note that there are regional variations both in styles available, and in their popularity. For example, the hot-water panel-radiator is popular in Europe for both domestic and commercial heating systems. In North America, variations on the finned-tube radiator are most popular. The panel radiator shown in *Figure 8.1* is manufactured in a range of heights, from 8 to 36 inches, and in lengths up to 8 feet.

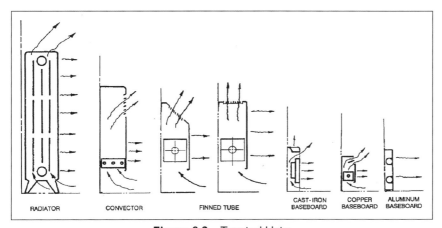

**Figure 8.2** Terminal Units

The radiator emits heat by both radiation and convection. The unit temperature is typically below 220°F and is considered 'low temperature' as far as radiation is concerned. In the final chapter of this book, we will review higher temperature radiant heaters and their specific characteristics and uses.

Starting at the left, we see the classic sectional radiator. Originally made from cast iron, there are now pressed-steel versions being manufactured. All of these terminal units are closed systems that heat the room-air as it contacts the heated coils.

The convector is a coil, mounted horizontally, at the bottom of a casing. The casing is open at the bottom and has louvers near, or in, the top. The coil heats the air, which becomes less dense and rises up the unit. The column of warm, less dense air causes a continuous flow over the coil, convecting heat from the unit. This warm air, rising in an enclosure, is called the "chimney effect," since it is most often experienced in the draft up a chimney. The taller the chimney, or in this case the taller the casing, the greater the draft through the unit, and the higher the output.

Convectors are typically used where medium output is required in a short length of wall.

The finned tube is similar to the convector, but the unit is long, and typically, runs around the perimeter of the building. The hot water enters one end and cools as it flows through the finned tube. If the fins on the tube are at a constant spacing, the output will fall as the water cools down. This drop in output can be offset, to some extent, by having sections of pipe with no fins at the hot end and also by changing the fin spacing along the tube.

Since the output occurs along the length of the unit, it nicely balances the heat loss through walls and windows, providing a thermally comfortable space without downdrafts. The construction is normally quite lightweight though, so if the finned tube is to be installed where someone may sit or stand upon it, a more robust version should be chosen. Some designs permit limited, manual adjustment to the output, accomplished by setting a flap damper in the unit.

The copper baseboard radiator is a small residential version of the finned tube. Cast iron baseboards have the advantage of being robust, however low output and substantial material make them less popular nowadays. Finally, the aluminum baseboard unit consists of pipes bonded to an aluminum sheet that emits almost all its heat by radiation, with a consequently low output.

These water heaters can all be controlled in two ways:

By varying the water flow
By varying the water supply temperature.

### *Varying the Water Flow*

Local zone control can be achieved by throttling the water flow. The simplest way to achieve this is with a self-contained control valve, mounted on the pipe. This valve contains a capsule of material that experiences large changes in volume, based on room temperature. As the temperature rises, the material expands and drives the valve closed. The valve settings are not marked with temperatures and it is a matter of trial-and-error to find the comfortable setting. A better, but more expensive, method of control is a wall thermostat and water control valve.

Control by modulating, or adjusting, the water flow works best when the load is high and the flow is high. For example, a finned tube, operating at low load with a low flow, will have almost full output just at the entry point of the water, but the water cools down to provide no output of heat at the far end. Both this issue and unnecessary pipe losses can be greatly reduced by modulating the water temperature.

### Varying the Water Temperature

The heat loss through a wall or window is proportional to the temperature difference across the wall or window. Thus, one can arrange a control system to increase the water temperature as the outside temperature falls, so that the heat output from the water will increase in step with the increase in heating load. This control system is called **outdoor reset**. In a simple outdoor reset system, the water flow temperature might be set to 180°F at the anticipated minimum outside design temperature, dropping to 70°F at an outside temperature of 70°F.

The output from the heater is not exactly linearly proportional to the water temperature. The actual output rises proportionately faster, the higher the temperature difference between heater and space. This disparity does not matter if the zone thermostat controls the zone temperature. Outdoor reset

- Minimizes uncontrolled heat loss from distribution piping.
- Improves zone control by keeping the zone flow control valves operating near full capacity.
- Achieves a more even temperature in the heaters, since the flow stays up.

Together, outdoor reset of water supply temperature and zone throttling provide excellent temperature control of hydronic systems.

### Meeting Ventilation Requirements

These hydronic heating systems do not provide any ventilation air from outside. When water systems are in use, ventilation requirements can be met in one of 3 ways:

Open windows
Window air conditioners
Separate ventilation systems with optional cooling.

**Open Windows**: Water systems are often used with occupant-controlled windows (opening windows) where the room depth is limited and the outdoor temperatures make it practical to open windows.

**Window Air conditioners**: One step up from heating and opening windows is heating and the window air-conditioner.

**Separate ventilation systems with optional cooling**: The alternative is to install a separate system to provide ventilation and, if needed, cooling. This is a very common design in cooler climates for two reasons. First, the water heating around the perimeter is very comfortable and, second, it means that the air system can be shut off when the building is unoccupied, leaving the heating operating and keeping the building warm. Many office buildings operate only five days a week, twelve hours a day, so the air system can be turned off for

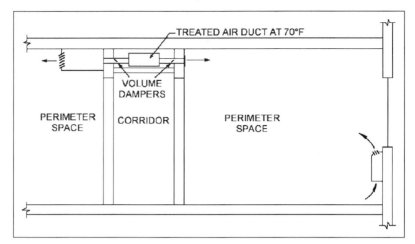

**Figure 8.3** Ventilation from a Separate Duct System

108 hours and only run 60 hours a week, saving 64% of the running hours of the ventilation system. *Figure 8.3* shows perimeter fan coils which provide heating and cooling plus a ventilation system using the corridor ceiling space for the ventilation supply duct.

The control of the hydronic heating system and ventilation/cooling system should be coordinated to avoid energy waste. Let us assume for a moment that each system has its own thermostat in each zone. If the heating thermostat is set warmer than the cooling thermostat, both systems will increase output until one is running flat out. Therefore, it is important to have a single thermostat controlling both the water heating system and the air-conditioning system. Ideally, this thermostat will have a **dead band**, which is a temperature range of, say, 2°F between turning off the cooling and turning on the heating.

In hot moist climates, the primary ventilation air must be supplied with a low moisture content to minimize mold problems. In addition, it is advantageous to keep the building pressure positive with respect to outside, so as to minimize local infiltration that might cause excessive moisture inside.

## 8.3 Panel Heating and Cooling

The floor or ceiling of the space can be used as the heater or cooler. A floor that uses the floor surface for heating is called a **radiant floor**.

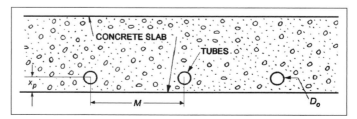

**Figure 8.4** Concrete Radiant Floor

The radiant floor is heated by small-bore plastic piping that snakes back and forth at even spacing over the entire area that requires heating. The output can be adjusted from area to area by adjusting the loop spacing, typically 6 to 18 inches, and circuiting the pipe loop. Typically the water is supplied first to the perimeter, to produce the higher output at the perimeter.

The acceptable floor surface temperature for occupants' feet limits the output. You may remember from Section 3.4, on human comfort, that ASHRAE Standard 55 limited the floor temperature to a range of 66–84°F for people wearing shoes who were not sitting on the floor. The maximum temperature limits the amount of heat that can be provided by a radiant floor.

Though radiant floors are often more expensive to install than other forms of heating, they can be very effective and economical to run, since they do not generate significant thermal stratification. As a result, the system is very comfortable and ideal for children and the elderly. Control is usually achieved by outdoor reset of water temperature and individual thermostats for each zone.

The system can also be installed in outside pavement by using an inhibited glycol (anti-freeze) mixture instead of plain water. This can be used to prevent icing of walkways, parking garage ramps and the floor of loading bays that are open to the weather.

Ceilings can also be used for heating and/or cooling. As noted in 3.4, when using ceilings for heating, care must be taken to avoid radiating too much heat onto occupants' heads. For ceilings down at 10 feet, the maximum temperature is 140°F. This maximum rises to 180°F at 18 feet ceiling height. When cooling, you circulate chilled water, instead of hot water through the ceiling panel pipe. The water temperature must be kept warm enough to ensure that condensation problems do not occur. The temperature difference between the ceiling panel and the space is quite limited. This limits the cooling capacity of the ceiling system and effectively limits its use to spaces that do not have high cooling loads.

Typically, a metal ceiling tile has a metal water pipe bonded to it, so that the whole surface becomes the heat emitter. There are many designs; one is shown in *Figure 8.5*.

The system has the advantage of taking up no floor or wall space and it collects no more dirt than a normal ceiling, making it very attractive for use in hospitals and other places that must be kept very clean.

## 8.4 Fan Coils

Up to now, the systems we have considered are passive (no moving parts) heating and cooling systems. We will now consider fan coils. As their name suggests, these units consist of a fan and a coil. Fan coils can be used for just heating or for both heating and cooling. In heating-only fan coils, the heating coil usually has fairly widely spaced fins so a lint filter is not critical. In dusty, linty environments, this may necessitate occasional vacuuming of the coil to remove lint buildup. Fan coils can be mounted against the wall at the ceiling. A typical fan-coil unit is illustrated in *Figure 8.6*.

When the fan-coil is used for heating, the hot water normally runs through the unit continuously. Some heat is emitted by natural convection, even when the fan is "off." When the thermostat switches the fan "on," full output is achieved. A thermostat within the unit works well in circulation areas, such as entrances and corridors, where temperature control is not critical, and

110    Fundamentals of HVAC

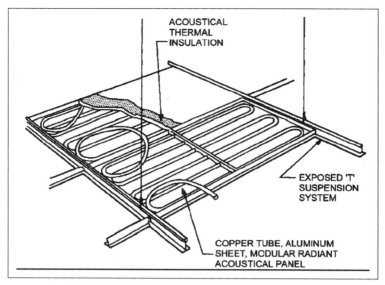

**Figure 8.5**   Example of Ceiling Radiant Panels

temperature differential is large. Generally, in occupied spaces, a room thermostat should be used to control the unit, to provide more accurate control.

Some units are provided with two or three speed controls for the fan, allowing adjustment in output of heat and generated noise. Many designers will choose a unit that is designed to run at middle speed, to minimize the noise

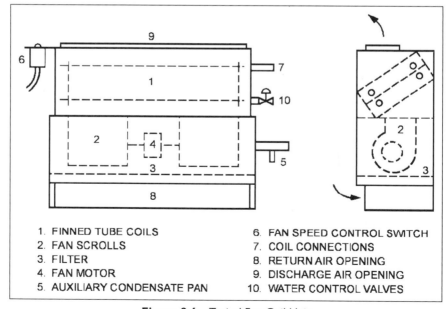

1. FINNED TUBE COILS
2. FAN SCROLLS
3. FILTER
4. FAN MOTOR
5. AUXILIARY CONDENSATE PAN
6. FAN SPEED CONTROL SWITCH
7. COIL CONNECTIONS
8. RETURN AIR OPENING
9. DISCHARGE AIR OPENING
10. WATER CONTROL VALVES

**Figure 8.6**   Typical Fan-Coil Unit

from the unit. Another way to minimize noise from the unit is to mount the unit in the ceiling space in the corridor and duct the air from the unit into the room.

**Hot-water fan coils**. These are an ideal method of providing heat to the high, sporadic, loads in entrances. In cold climates, if the outside door does not close, the unit can freeze, so it is wise to include a thermostat that prevents the fan from running if the outflow water temperature drops below 120°F. Fan-coils may be run on an outdoor-reset water system, but this limits their output and keeps the fan running more than if a constant, say 180°F, water temperature is supplied to the unit.

**Changeover system.** The same fan coil can be used for heating or for cooling, but with chilled water instead of hot water. This is called a changeover system. If a coil is used for cooling, it can become wet, due to condensation, and so it requires a condensate drain. The drain requires a slope of 1/8 inch per foot, to ensure that the condensate does not form a stagnant pool in the condensate pan. Failure to provide an adequate slope can result in mold growth and consequent indoor air quality, IAQ, problems. For ceiling-mounted units, providing an adequate slope for the drain can be a real challenge.

If the coil is designed to run dry, with no condensation, then a filter is not absolutely necessary. However, if the coil may run wet, it must be protected with a filter with efficiency minimum efficiency reporting value (MERV) of not less than 6 when rated in accordance with ANSI/ASHRAE Standard 52.2, to minimize lint and dust buildup on the coil. Both the filter and the drain require regular maintenance and therefore access to the unit must be available.

Timing is the challenge of changeover systems: when to change over from heating to cooling and vice versa. For manual changeover systems, the spring and fall can create real headaches for the operator. The system needs to be heating at night but cooling for the afternoon. The question for the operator is "What time should the change occur?" The challenge can be reduced if there is a ventilation system with temperature control. When it is cool outside, the ventilation air is supplied cool, thereby providing some cooling. When it is warm outside, the ventilation air is supplied warm and that will provide a little heating.

Generally, the operator will choose a day and change the system over, so that the spaces are either excessively warm in the afternoon or cool in the morning. The advent of computerized controls has enabled designers to include sophisticated automatic programs that deal with the changeover issue far more effectively than through manual operation.

**Four-Pipe system**: As an alternative design to a changeover system, the unit can include two coils, heating and cooling, each with its own water circuit. This is called a four-pipe system, since there are a total of four pipes serving the two coils. This system is more expensive to install but it is a more efficient system that completely avoids the problem of timing for change over from heating to cooling.

The four-pipe fan-coil system is ideal for places like hotels, where rooms may be unoccupied for long periods. The temperature can be allowed to drift well above or below the comfort level, since the fan-coil has enough output on full-speed to quickly bring the room to a comfortable temperature. Once the comfortable temperature is achieved, the occupant can turn the unit down to a slower speed so that the temperature is maintained with minimal fan noise.

## 8.5 Two Pipe Induction Systems

When air moves through a space with speed, additional air from the space is caught up in the flow, and moves with the flow of the air. When this occurs, the room air that is caught up in the flow is called **entrained air**, or **secondary air**.

The two-pipe induction system uses ventilation air at medium pressure to entrain room air across a coil that either heats or cools. The ventilation-air, called **primary air**, is supplied at medium pressure and discharged through an array of vertical-facing nozzles. The high-velocity air causes an entrained flow of room air over the coil and up through the unit, to discharge into the room. The flow of room air through the unit has little energy, so obstructing the inlet or the outlet with furniture, books etc. can seriously reduce the performance of the unit.

The coil in the induction unit is heated or cooled by water. For cooling, the coil should be designed to run dry, but it may run wet, so a condensate tray is normally necessary. In a hot, humid climate, to minimize the infiltration of moist air and reduce the likelihood of the coil running wet, the building pressure should be maintained positive. A lint filter should be provided to protect the coil. This filter will need to be changed regularly, so good access to the front of the unit is required.

The induction unit produces some noise due to the high nozzle velocity. This makes it less suitable for sleeping areas. The air noise is tone-free, though, and thus not annoying in most occupied spaces if silence is not a prerequisite.

The units are typically installed under a window, and when the air system is turned off the unit will provide some heat by natural convection, if hot water is flowing through the coil.

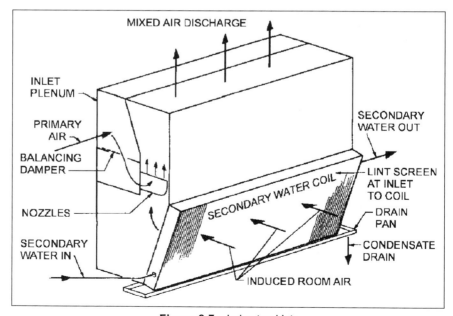

**Figure 8.7** Induction Unit

## 8.6 Water Source Heat Pumps

Water source heat pumps are reversible refrigeration units. The refrigeration circuit is the one we considered in Chapter 6, Figure 6-6 except that one coil is water cooled/heated instead of air cooled/heated. The heat pump can either transfer heat from water into the zone or extract heat from the zone and reject it into water. This ability finds two particular uses in building air conditioning:

   The use of heat from the ground
   The transfer of heat around a building.

### The Use of Heat from the Ground

There is a steady flow of heat from the core of the earth to the surface. As a result, a few feet below the surface, the ground temperature remains fairly steady. In cool climates, well below the frost line, this ground heat temperature may be only 40°F, but in the southern United States it reaches 70°F. This constant temperature can be utilized in two ways. Where there is groundwater available, two, properly distanced, wells can be dug and the water pumped up and through a heat pump. The heat pump can cool the water and heat the building or, in reverse, heat the water and cool the building.

Where the water is too corrosive to use, or not available, water filled coils of plastic pipe can be laid in the ground in horizontal or vertical arrays to absorb heat from, or dissipate heat into, the ground. This use of heat from the ground by a heat pump is commonly called a "ground-source heat pump."

The ground-source heat pump provides relatively economical heating or cooling using electricity. The ground-source heat pump has a much higher cooling efficiency than an air-cooler air-conditioning unit, making it very attractive in areas where the summer electricity price is very high or supply capacity is limited. In places where other fuels for heating are expensive, the ground-source heat pump can be very attractive.

### The Transfer of Heat Around a Building

The second use of heat pumps in building air conditioning is the water loop heat pump system. Here each zone is provided with one or more, heat pumps, connected to a water pipe loop around the building, see *Figure 8.9*. The water is circulated at 60°F to 90°F and the pipe is normally not insulated. Each zone heat pump uses the water to provide heating or cooling as required by that zone.

As you can see in *Figure 8.8*, there is a boiler to provide heating and a cooling tower to reject heat when the building has a net need for heating or cooling. The boiler, or tower, is used when required to maintain the circulation water within the set temperature limits. The system provides local heating or cooling at any time and each heat pump can be scheduled and controlled independently.

The question is: "why would anyone design a system that required so much equipment in a building?"

- In many buildings there are significant interior spaces that always require cooling, due to the heat from occupants, lighting, and equipment. This heat is put into the water loop and can then be used in exterior zones for heating.

- In addition, there are often times when the solar heat gain on the south side of a building requires zone cooling when the sun shines, while the north side of the building still requires heating.
- Lastly there are buildings with significant heat generation equipment, such as computer rooms, server racks, and telephone equipment, where the waste heat from these operations can be used to heat the rest of the building.

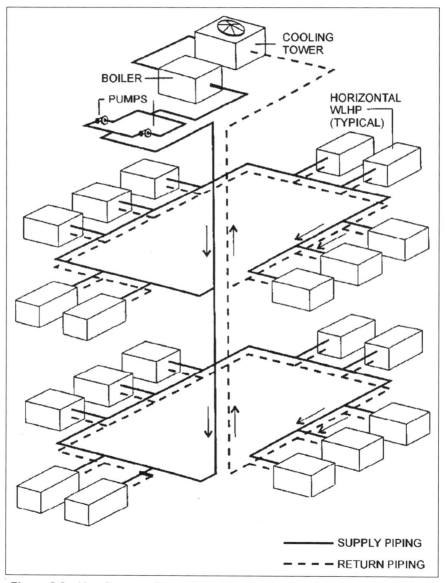

**Figure 8.8** Heat Recovery System Using Water-to-Air Heat Pumps in a Closed Loop

The heat pump units require regular filter changes to ensure that airflow is maintained, since they each include a direct expansion refrigeration circuit. In addition, the water circuiting must be designed to maintain a constant flow through the operating units, even when other units are removed for repair. This issue will come up again when we are discussing water piping in the next chapter, Chapter 9.

These closed loop systems are very effective in multiuse buildings, buildings with substantial core areas and heating loads, and buildings where occupancy is variable in both time and quantity. Examples include offices, hotels, commercial, colleges, and laboratories.

## The Next Step

Having considered a variety of hydronic systems in this chapter we will go on in Chapter 9 to consider the pumping, piping, balancing and control of water systems.

## Summary

This chapter has covered the more common hydronic systems used in air-conditioning buildings.

### 8.2 Natural Convection and Low Temperature Radiation Heating Systems

The very simplest water heating systems consist of pipes with hot water flowing through them. The output from a bare pipe is generally too low to be effective, so an extended surface is used to dissipate more heat. The radiator emits heat by both radiation and convection. These water heaters can all be controlled by varying the water flow or by varying the water supply temperature.

These hydronic heating systems do not provide any ventilation air from outside. When water systems are in use, ventilation requirements can be met by opening windows, window air conditioners, or separate ventilation systems with optional cooling.

### 8.3 Panel Heating and Cooling

Radiant floors use the floor surface for heating. Ceilings can also be used for heating and/or cooling. The system has the advantage of taking up no floor or wall space and it collects no more dirt than a normal ceiling.

### 8.4 Fan Coils

Fan coils can be used for just heating or for both heating and cooling. When the fan-coil is used for heating, the hot water normally runs through the unit continuously. Some heat is emitted by natural convection, even when the fan is off. When the thermostat switches the fan on, full output is achieved. Some units are provided with two, or three speed controls for the fan, allowing adjustment in output of heat and generated noise. Types of fan coils include: Hot-water fan coils, changeover systems, and four-pipe systems.

## 8.5 Two Pipe Induction Systems

The two-pipe induction system uses ventilation air at medium pressure to entrain room air across a coil that either heats or cools. The units are typically installed under a window, and when the air system is turned off, the unit will provide some heat by natural convection if hot water is flowing through the coil.

## 8.6 Water Source Heat Pumps

Water source heat pumps are refrigeration units that can either pump heat from water into the zone or extract heat from the zone and reject it into water. This ability finds two particular uses in building air conditioning:

1. The use of heat from the ground
2. The transfer of heat around a building.

## Bibliography

1. ASHRAE 2000 Systems and Equipment Handbook
2. ASHRAE 2001 Fundamentals Handbook

Chapter 9

# Hydronic System Architecture

## Contents of Chapter 9

Study Objectives of Chapter 9
9.1 Introduction
9.2 Steam
9.3 Water Systems
9.4 Hot Water
9.5 Chilled Water
9.6 Condenser Water
The Next Step
Summary
Bibliography

## Study Objectives of Chapter 9

Chapter 9 introduces you to the various hydronic distribution systems and some of their characteristics. Because this chapter is in a fundamentals course, we will not be developing detailed design information. For detailed information about water systems, you can take the ASHRAE Course, *Fundamentals of Water Systems*[1].

When you have completed this chapter, you should be familiar with:

**Steam systems**: The general operation and some of the advantages and disadvantages of steam distribution systems.
**Hot water heating systems**: The main piping-layout options, pumping requirements and characteristics
**Chilled water systems**: The popular piping arrangements and characteristics
**Open water systems**: The behavior of a condenser, condenser requirements, and cooling tower operation

## 9.1 Introduction

In previous chapters, we have considered a variety of systems that need a source of heat or cooling to operate. Many of these systems use water or steam for this source. This chapter will introduce you to the basic layout options for heating and cooling piping arrangements that distribute water or steam,

118    Fundamentals of HVAC

**hydronic circuits**. It will also provide a brief discussion of the differences in their hydronic characteristics.

In each case, a flow of water or steam is distributed from a either a central boiler or a **chiller**, the refrigeration equipment used to produce chilled water, to the hydronic circuits. The hydronic circuits circulate the water or steam through the building, where it loses or gains heat before returning to be re-heated or re-cooled.

The water or steam is treated with chemicals to inhibit corrosion and bacterial growth in the system.

## 9.2 Steam

Steam results from boiling water. As the water boils, it takes up latent heat of vaporization and expands to about 1,600 times its original volume at atmospheric pressure. Steam is a gas, and in a vessel, it quickly expands to fill the space available at a constant pressure throughout the vessel. In this case, the relevant space is the boiler(s) and the pipe that runs from the boiler and around the building. The pipe rapidly fills with steam, and the pressure is virtually the same from end to end under no-flow conditions. As flow increases, there is a pressure drop due to friction against the pipe wall and due to the energy needed to produce flow.

When the steam gives up its latent heat of evaporation in an end-use device, such as a coil, fan coil, or radiator, it condenses back to water, and the water is called "**condensate.**" This condensate is removed from the steam system by means of a "**steam trap**." A steam trap is so-named because it traps the steam while allowing the condensate out of the higher-pressure steam system into the lower-pressure condensate return pipe.

Traps are typically thermostatic or float operated.

**Thermostatic Trap**: In the thermostatic trap, a bellows is used to hold the trap exit closed when heated by steam. The bellows is filled with a fluid that boils at just below the steam temperature. When the trap fills with air or condensate, the temperature drops and the bellows contract, letting the air or condensate flow out. As soon as the air or condensate is expelled and the trap fills with steam, the heated bellows expands, trapping the steam.

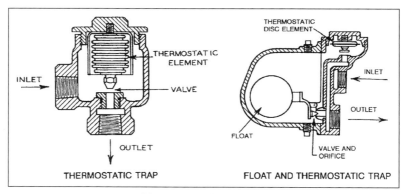

**Figure 9.1**  Steam Traps

**Float and Thermostatic Trap**: This versatile trap uses the much higher density of condensate to lift a float to open the trap and release the large quantities of condensate produced under startup and high-load periods. When filling the system, large volumes of air must be vented. The thermostatic element works well for this function. During operation at low loads, the float functions well to drain the slow accumulation of condensate. In most systems, the condensate is gravity-piped to a condensate collection tank, before being intermittently pumped back to the boiler makeup tank. Due to the much smaller volume of condensate, the condensate return piping is smaller in diameter than the steam supply pipe.

Regardless of which trap is used, the returned condensate and any required makeup treated water are pumped into the boiler to be boiled into steam again. The initial water fill and all water added to the steam boiler must be treated to remove oxygen and harmful chemicals that could cause serious corrosion in the boiler and pipe work. The addition of these chemicals means that the water in the system is **not potable**, not suitable for human consumption. As a result, the steam from the heating distribution system is unsuitable for injecting into the air for humidification. However, the heating steam can be used to indirectly evaporate potable water for humidification, where required.

Because steam has low density and the ability to move itself throughout the system, it is ideal for use in tall buildings. The steam makes its own way to where it is needed and gravity brings the condensate back down again.

*Figure 9.2* shows the main components of a small steam system. The condensate is pumped into the boiler where it is boiled into steam. The steam expands down the main and into any heater that has an open valve. As the heater gives off heat, the steam condenses. The condensate collects at the bottom of the heater and is drained away by the trap.

Steam systems are divided into two categories: low-pressure systems and high-pressure systems. **Low-pressure systems** operate at no more than 15 **"psig"**, meaning no more than 15 pounds-per-square-inch pressure higher than the local atmospheric pressure. **High-pressure systems** operate above 15 psig.

## Safety Issues

In order to maintain the system pressure, the boiler output needs to be continuously balanced with the load. Because steam has the capacity to expand at high velocity in all directions, a poor boiler operation can cause an accident.

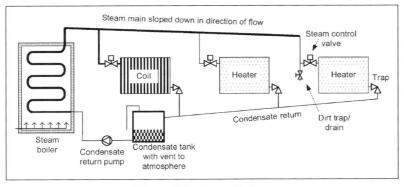

**Figure 9.2** Steam System

The requirements for boiler operations on low-pressure systems are very much less stringent compared to high-pressure systems. Early in the twentieth century, there were numerous boiler explosions. As a result, the American Society of Mechanical Engineers wrote strict codes for the manufacture of steam boilers and associated piping and equipment. Those codes have drastically reduced the number of failures in North America.

The local pressure vessel regulations are relatively rigorously enforced in most countries. The rules and regulations for both manufacture and operation vary substantially in different countries, so having local information is always a high priority when you are designing or operating a steam pressure system.

Steam systems need to be installed carefully, maintaining a downward slope of 1 in 500 to avoid condensate collecting, called **ponding**, in the steam pipe. If condensate ponds in the steam pipe and the steam flow increases significantly, a slug of condensate can be lifted and carried by the steam at very high velocity until it reaches a bend or other obstruction. The slug of water can attain a high momentum and may break the joint or valve. Not only can the pipe be ruptured, but as soon as the pipe is ruptured, the steam is free to escape and can easily burn, or kill, anyone in the area.

The advantages of steam are

Very high heat transfer.
No need for supply pumps.
Easy to add loads because the system adjusts to balance the loads.

These systems are much less popular than they used to be, but they are still an attractive choice for distribution of large amounts of heat around numerous or high buildings.

## 9.3 Water Systems

Water systems are more commonly used for heating than are steam systems. The advantages of water over steam include the fact that water is safer and more controllable than steam.

Water is safer because the system pressure is not determined by continuously balancing the boiler output with load, and because water does not have the capacity to expand at high velocity in all directions.

Water is more controllable for heating since the water temperature can easily be changed to modify the heat transfer.

### Water System Design Issues: Pipe Construction

Water for heating and cooling is transferred in pipes that are generally made of steel, copper or iron. Steel is normally a less expensive material and is most popular for sizes over 1 inch. Copper is a more expensive material but it is very popular at 1 inch and narrower, due to its ease of installation. Long runs with few fittings favor steel, while the more complex connections to equipment favor the easy installation of copper.

### Water System Design Issues: Pipe Distribution

Heating or cooling water can be piped around a building in two ways, either "**direct return**" or "**reverse return.**" The direct return is diagrammed in *Figure 9.3*.

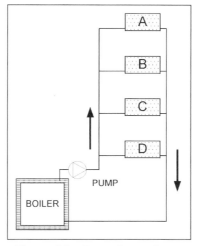

**Figure 9.3**  Direct Return Piping

The simple circuit in *Figure 9.3* consists of a boiler; four identical heaters A, B, C, D; a pump to drive the water round the circuit; and interconnecting pipes. When the pump is running, water will flow from the boiler to each heater, through the heater, and back to the pump, to be pumped around the circuit again.

There is friction to the water flowing through the pipes and the water favors the path of least resistance. The circuit: **boiler → pump → heater D → boiler**, is much shorter than the circuit: **boiler → pump → heater A → boiler**. As a result more water will flow through heater D than through heater A.

In order to have the same flow through all the heaters, extra resistance has to be added to heaters B, C and D. Adding balancing valves, as shown in *Figure 9.4*, makes this possible.

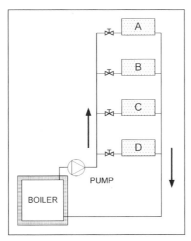

**Figure 9.4**  Direct Return Piping with Balancing Valves

After the system has been installed, a balancing contractor will adjust the balancing valves to create an equal flow through heaters A and B, then an equal flow through heaters A and C and finally an equal flow through heaters A and D. This simple, step-by-step, procedure will produce the highest balanced set of flows for the four heaters.

The total flow may be more or less than design, but the flows will be equal. If the flow is more than required, it is possible, but difficult, to go back and rebalance to a specific lower flow.

In practice, a single balancing valve in the main loop, often between the pump and boiler, can be used to reduce the total flow. As the total flow is reduced, the flow in each heater will reduce in the same proportion. This circuit works well, once it has been balanced. On most systems, a valve is installed on each side of heaters so that the heater can be valved-off and repaired without having to shut down and drain the whole system.

Let us now imagine that one of the heaters failed and in the process of removing it, the balancing valve is closed. When the heater has been replaced the question is "How much should the balancing valve be opened?" Did anyone take note of the valve position before it was moved? If not, the balancing valve will likely be left fully open. The system may work satisfactorily with the balance valve open, or, it may not. This problem of being dependent on balancing valves can largely be overcome by using a different piping arrangement, the reverse return as shown in *Figure 9.5*. Here the pipe length for the flow loop **boiler** → **pump** → **heater** → **boiler** is the same for all heaters. Verify this for yourself by tracing the water path through heater D and then the path through heater A. As a result, the flow will be the same in each heater; the piping is self-balancing.

The reverse-return piping costs more due to the additional return length of pipe. There are cases where the flow is critical, for example, direct expansion refrigeration heat pumps. In this case, the additional cost of reverse return piping is worthwhile. The maintenance staff only needs to fully open the valves to a unit to know it has full flow.

In circuits where exact balance is not critical, a system with direct return and balancing valves is a good choice.

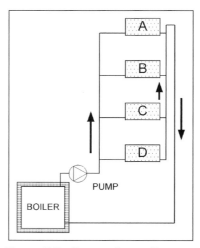

**Figure 9.5** Reverse Return Piping

Having considered the two main piping arrangements let us now go on to the flow of water and pumps.

### *Water System Design Issues: Flow*

The resistance to water flow in pipes, called the **head**, is dependent on several factors including surface roughness, turbulence, and pipe size. When we design a system, we calculate the expected resistance for the design flow in each part of the circuit. The sum of the resistances gives the total resistance, or **system head**.

Under normal flow rates, the resistance rises by a factor of 1.85 to 1.9 as the flow rises (flow$^{1.85}$ to flow$^{1.9}$). Doubling the flow increases the resistance about three and a half times.

The actual **head loss** in pipes is normally read from tables, to avoid repetitive complex calculations. Based on this table data and the knowledge that the head is proportional to flow$^{1.85}$ we can plot the **system curve** of flow or capacity, versus head.

Pump manufacturers test their pumps to establish what flows the pump generates at a range of heads. At a particular pump speed, measured in revolutions per minute, **rpm**, they will measure the head with no flow, and again at increasing flows, or capacities. They can then plot the pump head against flow or capacity to produce a **pump curve**. A pump curve and calculated system curve are shown in *Figure 9.6*.

The pump curve in this figure shows a peak head of 90 feet with no flow that gradually drops to about 65 feet at 74 gallons per minute, **gpm**, where it crosses the calculated system curve. If the design calculations were correct, the **operating point** for this pump will be at the intersection of the two curves.

In practice, the system curve often turns out to be higher or lower than the calculated design. The effects of this, and remedies for it, are covered in the course ASHRAE *Fundamentals of Water Systems*.

The layout of piping in a building is very dependent on load locations and where pipe access is available. *Figure 8.8*, in the last chapter, showed a single riser

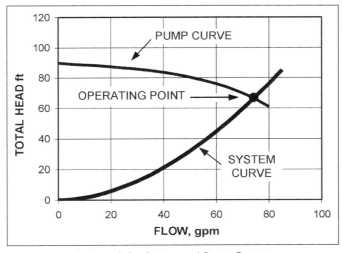

**Figure 9.6** System and Pump Curves

124    Fundamentals of HVAC

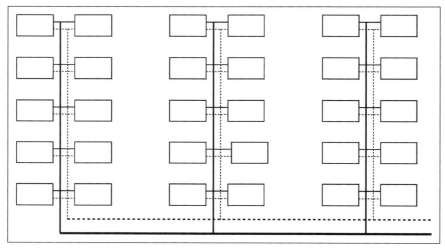

**Figure 9.7**  Multiple Risers

in the building with a reverse return loop around every floor. This works well for heat pumps mounted in the ceiling, with the pipes running in the ceiling.

Conversely, it often does **not** work very well for equipment, such as radiators, fan coils and induction units, mounted near the floor at the perimeter of the building. For these, multiple risers around the building may be a better solution, as shown in *Figure 9.7*.

Having introduced piping layouts and pumps let us go on to consider the three main types of water circuits and some of their characteristics.

## 9.4 Hot Water

Within buildings, hot water is the fluid that is most commonly used for heat-distribution. The amount of heat that is transferred is proportional to the temperature difference between supply and return. Maximizing the supply-return temperature difference minimizes the water quantity and pipe size requirements. Unfortunately, the economy of smaller water quantities with a high temperature difference creates a need for larger, and more costly, heaters and heat exchangers. The design challenge is thus to find the best balance between cost to install and cost to operate.

For general use, in buildings where the public may touch the pipes, the normal operating supply temperature is 180°F. In the past, return temperatures were 160°F, but temperatures of 150°F, or even 140°F, are now often used for overall operating economy. Systems can also be designed to operate with a 180°F flow, except under peak load conditions. Peak load conditions hardly ever occur, but if they do, then the flow temperature can be raised as high as 200°F.

These systems can operate at very low pressure, since the only requirement is that the pipes remain full. For working temperatures above 200°F, at sea level, systems must be pressurized to avoid the possibility of the water boiling.

As discussed in the previous chapter, radiant floors operate with a maximum surface temperature of 84°F. They need heating water at 120°F or less, much cooler than 180°F. This can be achieved by mixing cool return water with the 180°F water to provide a supply to the floor at 120°F or less. Alternatively, and very fuel efficiently, they can be supplied from a condensing boiler or ground source heat pump, both of which have a maximum flow temperature of about 120°F.

For distribution between buildings, higher temperatures—up to 450°F—can be used. The high temperature hot water is passed through a heat exchanger in each building to provide the, typically, 180°F water for distribution around the building and for heating domestic hot water.

Pipes should be insulated to avoid **wasteful** heat loss. Thus pipes in the boiler room should be insulated, but pipes in a zone that is feeding a radiator may not need to be insulated, since the heat loss just adds to the radiator output. However, if a pipe presents an exposed surface that could cause a burn, insulation should be used.

Insulation thickness should take into account the temperature difference between the water and surroundings. Thus, rather thicker insulation should be used on pipes that run outside a building than inside the building.

## 9.4.1 Energy Efficiency in Hot Water Systems

There are many ways to control and increase energy efficiency in the hot water systems. The control method that we will discuss is the **outdoor reset**, a common control strategy that takes advantage of the temperature differential between the cold outside and the warm inside the building to adjust the heat output. Then we will consider pumps and the energy savings that we can obtain through reducing the flow in hot water systems:

The heat loss from a building in cold weather is proportional to the difference between the temperature inside the building and the temperature outside the building. Similarly, the heat output from a convection heater is roughly proportional to the difference in between the space temperature and the heating-supply-water temperature. Outdoor reset makes combined use of these two relationships by adjusting the heating-water temperature with changes in outdoor temperature. With the correct schedule, the water flow remains constant and the heat output just balances the building heat loss.

This outdoor reset system has advantages, but it does mean that the heating water flow is 100% all through the heating season. This continuous full flow involves a significant pumping cost.

In the last section we noted that the head is proportional to the flow$^{1.85}$. The pumping power is proportional to the head, times the flow. So, doubling the flow requires

$2(2^{1.85}) = 7.2$ times the power!

Here is an incentive to reduce flow. If, instead of modulating the water temperature, it remained constant at, say 180°F, and the flow was varied by thermostatic valves, the required flow would be much less than 100% most of the time. In fact, since most heating systems are oversized, the flow would never reach 100%. However, as soon as the flow varies, we need a method of varying the pump capacity.

126    Fundamentals of HVAC

In the following sections, we will consider two methods of varying pump capacity:

1. Varying pump speed
2. Using pumps in parallel

**Varying Pump Speed** Variable speed drives are now readily available and can be used to adjust pump speed according to load. The pump curve remains the same shape, but shrinks as the speed reduces. Typical pump curves for various speeds are shown in *Figure 9.8*.

The arrows in the figure indicate that the head is about 25% at 50% speed and 50% flow, while the power consumption is about 10% at 50% flow.

The figure also shows the **pump shaft power**, which is the power used by the pump, without consideration of any bearing or motor inefficiencies. Since motor efficiency generally drops significantly at low speeds, the overall reduction in power is much less than the figure indicates at low flows.

**Pumps In Parallel** Another way to reduce flow is to use two identical pumps in parallel. Each pump experiences the same head, and their flows add to equal the system flow. A check valve is included with each pump, so that when only one pump is running, the water cannot flow backwards through the pump that is "off." The piping arrangement is shown in *Figure 9.9*.

With both pumps running, the design flow is at the system operating point. When one pump is shut off, the flow and head drop to the single pump curve as shown in *Figure 9.8*. This flow is between 70% and 80% of full flow, depending on pump design. **Note** that the power required by the single pump is slightly higher when running on its own and the motor must be sized for this duty.

The use of parallel pumps for a heating system has two advantages: First, it produces a substantial reduction in energy use for all the hours the system is using only one pump; second, it provides automatic stand-by to at least 70% duty when one of the pumps fails.

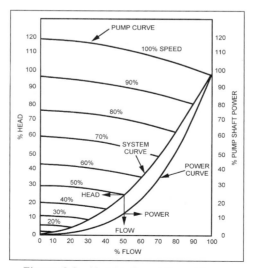

**Figure 9.8**   Variable Speed Pump Curves

Hydronic System Architecture    127

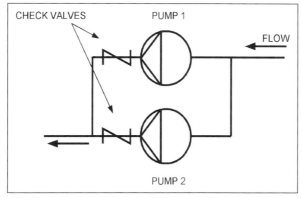

**Figure 9.9**   Pumps in Parallel

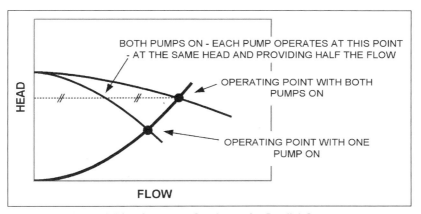

**Figure 9.10**   Operating Conditions for Parallel Operation

## 9.5 Chilled Water

Chilled water typically has a supply temperature of between 42°F and 48°F. Historically, the return temperature was often chosen to be 10°F above the flow temperature. With the higher cost of fuel and the concern over energy usage, it is usually cost effective to design for a higher difference of 15°F or even 20°F. The higher return temperatures require larger coils, and create challenges when high dehumidification is required.

On the other hand, doubling the temperature difference halves the volume flow, and, consequently, reduces the purchase cost of piping and pumps, as well as substantially reducing ongoing pumping power costs.

With a flow temperature in the range 42°F to 48°F, the piping must be insulated to reduce heat gain and avoid condensation. The insulation requires a moisture barrier on the outside to prevent condensation on the pipe.

**Chillers**, the refrigeration equipment used to produce chilled water, mostly use a direct expansion evaporator. Therefore, the flow must be maintained fairly constant to prevent the possibility of freezing the water. The chiller

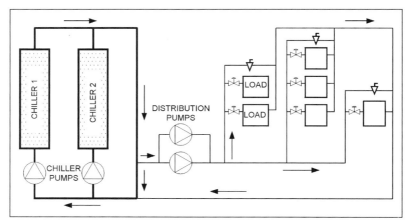

**Figure 9.11**  Chiller System with Decoupled Flows

requires constant flow but it would be both convenient and economical to have variable flow to the loads. To achieve this, the chiller and loads can be hydraulically "**decoupled**." Decoupled, in this context, means that the flows in the chiller circuit do not influence flows in the load circuit. Conversely, changes in the flows in the load circuit do not affect the chiller circuit.

A diagram of two chillers and loads is shown in *Figure 9.11*. The two chillers are piped in parallel in their own independent pipe loop, shown bold in the Figure. The chiller loop can run even if the distribution pumps are off. Similarly, the distribution loop can run with the chiller pumps off. The short section of shared pipe allows both loops to operate independently of each other, decoupled.

Each chiller has a pump that runs when the chiller runs, producing a chiller-circuit flow of 50% or 100%. The flow in the cooling-loads circuit is dependent on the distribution pumps and whether the valves are fully open or throttling (reducing) the flow. If the chiller flow is higher than the coil circuit, water will flow through the short common section of pipe as the excess chiller water flows round and round the chiller loop. If the chiller flow is less than the coil circuit flow, than some coil return water will flow through the short common section of pipe and mix with the chilled water. When this happens, a flow or temperature sensor will detect it and start another chiller.

The loads in *Figures 9.11* and *9.12* are shown as having two way valves which have no flow when they are closed. If all the valves were to close, the pump would be pumping against a closed circuit. To avoid problems occurring when this happens, a bypass valve is shown across the end of each branch circuit to allow a minimum flow under all conditions.

The arrangement in *Figure 9.11*, with distribution pumps serving all loads, requires these pumps to run regardless of the load. On projects where sections of load may be shut down while others are running, a "**distributed**" pumping arrangement may be more efficient. In *Figure 9.12* each secondary loop has its own pump, which is sized to deal with its own loop resistance and the main loop resistance. This system allows pumps 1, 2, and 3 to be run independently, when necessary, to serve their own loads.

The development of economical and sophisticated computer control and affordable variable speed drives, now enables designers to organize piping and

Hydronic System Architecture    129

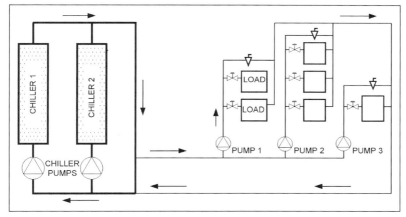

**Figure 9.12**  Distributed Secondary Pumping

pumping systems that really match need to power, compared to the historical situation where the system used full pump power whenever the system was "on."

## 9.6 Condenser Water

Condenser water is water that flows through the condenser of a chiller to cool the refrigerant. Condenser water from a chiller typically leaves the chiller at 95°F and returns from the **cooling tower** at 85°F or cooler. The cooling tower is a device that is used for evaporative cooling of water.

In *Figure 9.13* the hot, 95°F, water from the chiller condenser flows in at the top. It is then sprayed, or dripped, over fill, before collecting in the tray at the

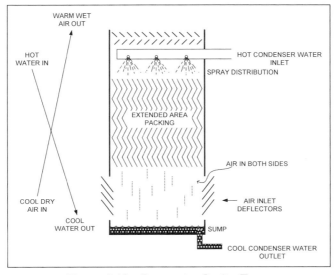

**Figure 9.13**  Evaporative Cooling Tower

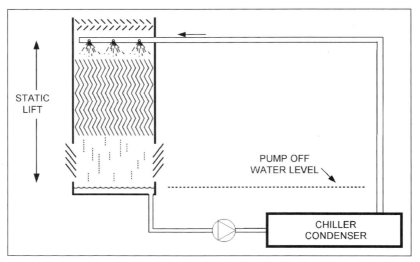

**Figure 9.14** Open Water Circuit

bottom. Air enters the lower part of the tower and rises through the tower, evaporating moisture and being cooled in the process, before exiting at the top.

We will consider cooling towers in more detail in the next chapter, but the tower has a hydraulic characteristic that we will cover here. The water has two open surfaces, the one at the top sprays and the other at the sump surface. This is an **open-water system**. An open-water system is one with **two**, or more open water surfaces. A **closed-water system** has only one water surface.

*Figure 9.14* shows an outline elevation of the complete cooling tower and chiller condenser water circuit. The water loop has two water surfaces, one at the top water sprays and one below at the sump water surface. When the pump is "off," the water will drain down to an equal level in the tower sump and in the pipe riser, as indicated by the horizontal dotted line in *Figure 9.14*. When the pump starts, it first has to lift the water up the vertical pipe before it can circulate it. The distance that the pump has to lift the water is called the **"static lift."** Once running, the pump has to provide the power to overcome both the static lift and the head, to overcome friction, to maintain the water flow.

*Figure 9.15* shows a closed water circuit. It is shown with one water surface open to the atmosphere. Whether the pump runs or not, the water level stays constant. When the pump starts, it only has to overcome friction to establish and maintain the water flow. When the pump stops, the flow stops, but there is no change in the water level in the tank. The open surface is required to allow for expansion and contraction as the water temperature changes during operation. In larger systems and most North American systems, the one open water surface is in a closed tank of compressed air rather than open to atmosphere, as is common in other parts of the world.

The cooling tower provides maintenance challenges. It contains warm water and dust, so it easily supports the multiplication of the potentially lethal bacteria, legionella.

We will return to cooling towers, their design, interconnection and operation when we discus central plants in the next chapter.

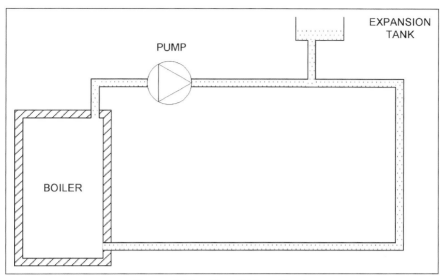

**Figure 9.15** Closed Water Circuit

## The Next Step

This chapter has covered hydronics architecture, specifically the piping systems for steam, hot water, chilled water and condenser water. In Chapter 10 we are going to consider the central plant boilers, chillers and cooling towers that produce the sources of steam and water at various temperatures.

## Summary

In this chapter, we covered hydronics systems, systems involving the flow of steam or water to transfer heat or cooling from one place to another.

### 9.2 Steam Systems

Principal ideas of this section include: how steam is used; its behavior as a gas and how it condenses as it gives up its latent heat; how the resultant condensate is drained out of the steam pipes by traps and then returned to the boiler, to be boiled into steam again.

### 9.3 Water Systems

In this section we described water systems and the economical direct arrangement and the more costly, but largely self-balancing, reverse-return piping arrangement. Once a system has been designed, the design flow and head are known and can be plotted on the same graph as the pump curve, to find the expected operating condition.

### 9.4 Hot Water Systems

From general water systems, we moved into hot water systems. The use, and energy savings of variable speed pumps was introduced. This was followed by a discussion of how two pumps in parallel can be used to provide reduced energy consumption for most of the heating season, as well as substantial, automatically-available, stand-by capacity should a pump fail.

### 9.5 Chilled Water

Because chilled water systems need constant water flow through the chiller evaporator, the economies of variable flow can be achieved through decoupled and distributed piping arrangements.

### 9.6 Cooling Towers

Cooling towers were described as well as the difference between open and closed water systems. The hot water and chilled water circuits are normally closed systems, but the cooling tower is an open system. The open system has a modified design requirement, since the pump must not only overcome the friction, head, to flow around the circuit, but must also provide lift to raise the water from the balance point to the highest point in the system.

## Bibliography

1. ASHRAE Course Fundamentals of Water System Design
2. 2004 ASHRAE Handbook Systems and Equipment
3. 2001 ASHRAE Handbook Fundamentals

Chapter 10

# Central Plants

## Contents of Chapter 10

Study Objectives of Chapter 10
10.1 Introduction
10.2 Central Plant Versus Local Plant
10.3 Boilers
10.4 Chillers
10.5 Cooling Towers
The Next Step
Summary
Bibliography

## Study Objectives of Chapter 10

In the last chapters we have discussed various air-conditioning systems and the fact that heating and cooling can be provided from a central plant by means of hot water, steam, and chilled water. In this chapter we will consider central plants. We will start with some general considerations about what they produce, their advantages, and their disadvantages. After studying the chapter, you should be able to:

Discuss some advantages and disadvantages of central plants.
Identify the main types of boiler and sketch a twin boiler circuit.
Describe the operation of chillers, and be able to sketch a dual chiller installation with primary only, and primary-secondary chilled water circuits.
Understand the operation of cooling towers, what affects their performance and what regular maintenance is required for safe and reliable operation.

## 10.1 Introduction

Central plants, for this course, include boilers, producing steam or hot water, and chillers, producing chilled water. These pieces of equipment can satisfy the heating and cooling requirements for a complete building. In a central plant, the boilers and chillers are located in a single space in the building, and their output is piped to all the various air-conditioning units and systems in the building. They are used in all types of larger buildings. Their initial cost is often higher than packaged units and they require installation floor area as well as space through the buildings for distribution pipes. Central plants generally require less maintenance than numerous smaller package systems and the equipment usually has a longer life.

This central plant concept can be extended to provide heating and cooling to many buildings on a campus or part of a town. The equipment for these larger systems is often housed in a separate building which reduces, or avoids, noise and safety issues.

We will be discussing some of the advantages and disadvantages of central plants and then we will go on to consider the main items of equipment found in central plants: boilers, chillers and cooling towers.

**Boilers** are pressure vessels and their installation and operation are strictly prescribed by codes. Their general construction, operation, and main safety features will be discussed.

**Chillers** come in a huge range of sizes and types and we will briefly introduce them. We will discuss their particular requirements for chilled water piping and specialized control.

The job of the chiller is to remove heat from the chilled water and reject it to the condenser. The condensers are often water-cooled. The cooling water is called "**condenser water**." The condenser water flows to a cooling tower, where it is cooled before it returns to the chiller to be heated once again. This will be discussed in detail in 10.4.

**Cooling towers** are devices used to cool water by evaporation. Water is sprayed or dripped over material with a large surface area, while outdoor air is drawn through. Some water evaporates, cooling the bulk of the water before it returns to the chiller.

## 10.2 Central Plant Versus Local Plant in a Building

There is no rule about when a central plant is the right answer or when distributed packages or systems should be used. Circumstances differ from project to project, and location to location. The good designer will assess each project on the merits of that situation and involve the client in making the most suitable choice for the project.

In this section we are going to consider, in a general way, some of the technical issues that can influence the choice. We are not going to consider the internal politics that can have major influences and costs in time and money. In addition to politics, the availability of money for installation versus operating costs can have a major impact on system choices. For minimum installation-cost, the package approach usually wins.

Here are some true statements in favor of central plants. Read them. Can you think of a reason why each one of them might, in some circumstances, be wrong, or irrelevant? Write down your suggested reason.

> "It is easy to have someone watching the plant if it is all in one place."
> "The large central plant equipment is always much more efficient than small local plant."
> "The endless cost of local plant replacement makes it uneconomic compared to a main central plant."

It is alright if you did not think of reasons, but do be aware that technology has changed over the last half century and you should think about whether categorical statements or "rules-of-thumb" are correct or relevant in your particular situation. You cannot go against the laws of physics, but there are many more ways of doing things than there were.

Let us consider each of the above statements in turn.

*"It is easy to have someone watching the plant if it is all in one place."*

This statement is true if visual inspection of the plant is useful. A hundred years ago, the look and sound of the plant were the best, and only, indicators of performance. Operators "knew their plant" and almost intuitively knew when things needed attention. Now, in the 21st century, plant is much more complex and we have excellent monitoring equipment available at a reasonable price. The information from those monitors can be instantly, and remotely, available. So instead of paying someone to physically watch the central plant, the building owner can pay someone to monitor the performance of, not just the central plant, but all the plant, regardless of where it is located in the buildings. Now, using the internet, many buildings can be monitored from anywhere in the world with fast and reliable internet service.

The second statement, *"The large central plant equipment is always more efficient than small local plant,"* is generally true but not always relevant. For example, an apartment building might have a large central boiler that provides both hot water for heating, and domestic hot water. In winter this is an efficient system. However throughout the summer the boiler will be running sporadically at very low load. It will take a considerable amount of energy to heat up the boiler before it starts to heat the domestic water, and this heat will dissipate to atmosphere before it is called on to heat the water again—very inefficient. The unit has a high efficiency at full load but when its efficiency is averaged over the year, "**seasonal efficiency**," may be surprisingly low.

In this situation, it may be beneficial to install a series of small hot-water heaters for the domestic hot water, although they are not as efficient as the main boiler at full load. Their advantage is that they only run when needed and have low standby losses.

The last statement *"The endless cost of local plant replacement makes it uneconomic compared to a main central plant."* is also true in some cases, but definitely not in other cases. In many organizations, replacement of smaller pieces of equipment are paid for as part of the maintenance operations' budget. On the other hand, major plant replacements are paid for out of a separate 'capital' fund. From the point-of-view of the maintenance managers, small, local plant is an endless expense to their maintenance budget, while other budgets fund large, central-plant replacements from the capital account. When it comes to new facilities, the maintenance managers in this situation are likely to be biased against small, packaged-plant equipment, because its replacement costs will all fall on their maintenance budget.

Let us go back to the reasons you wrote down as to why these three statements about central plant might be wrong. Are you still comfortable with them and can you think of others?

This section has deliberately been encouraging you to think about the some of the pros and cons of central plants. Now let us consider three other advantages.

1. "It is so much easier to maintain a high standard of operation and maintenance of a few large units in a single place, instead of lots of little packages all over the site."

Plant operators know that having complete information about the plant, all the tools in one place, space to work, and protection from the weather, all make central plant maintenance very attractive.

2. "Trying to optimize many package units is really difficult compared to the two identical chillers and boilers in our central plant."

A few central pieces of equipment can be monitored relatively easily and adjusted by the maintenance staff. When there are many units all over the building, it becomes difficult to remember which one is which and their individual quirks and characteristics.

3. "Heat recovery from central plant chillers and boilers is financially worth while."

Heat recovery is the recovery of heat that would otherwise have gone to waste. For example, the chiller absorbs heat from the chilled water and rejects it through the condenser to atmosphere. In a hospital with substantial hot water loads, some of this waste heat could be used to preheat the domestic hot water and perhaps to heat the air-conditioning reheat coils.

In a similar way, additional heat can be recovered from boiler flue gases by means of a **recuperator**. This is a device consisting of water sprays in a corrosion resistant section of flue. The water heats to around 120°F and is pumped through a water-to-water heat exchanger to provide water at about 115°F. This water can be used in an oversized coil for preheating outdoor air.

Both the heat-recovery from the chillers and recuperator-heat from the boilers are examples of the improved energy efficiency that is often not economically feasible on the smaller distributed-packaged equipment.

## 10.3 Boilers

Boilers are pressure vessels used to produce steam or hot water. They are different from **furnaces**, a term usually used to refer to air heaters of any size. Boilers come in a vast range of types and sizes.

The critical design factor is pressure. Boilers are fitted with safety valves that release the steam or water if the pressure rises significantly above the design pressure. The safety-equipment requirement and staff-monitoring requirements are far less stringent for low-pressure boilers, so there is a significant incentive to use low-pressure except where high pressure is needed, or more economic.

A "low-pressure" steam boiler operates at a pressure of no more than 15 pounds per square inch, **15 psig**, more than the local atmospheric pressure. This means 15 psig as measured by a **gauge** exposed to the local atmospheric pressure. In comparison, "low-pressure" hot water boilers are allowed up to 160 psig. There is a good reason for the extreme difference in allowable pressure:

> When a steam boiler fails, the effect can be catastrophic: as the steam expands uncontrollably, it is like a bomb going off. In comparison, when a hot water system bursts, the hot water pours out, but there is no explosive blast like there is with steam. For this reason, "low-pressure" hot water boilers are allowed up to the higher pressure of 160 psig.

Boilers and system components are regulated by codes. These codes are generally written, and updated, by practitioners in their geographic area. The main codes in North America are those issued by the American Society of Mechanical Engineers (ASME) *Boiler and Pressure Vessel Code* while the European Community has their own, and in many areas, much less demanding set of codes. It is therefore critical that a designer or operator knows the local code requirements, since their experience from one place may not be relevant in another jurisdiction.

## Boiler Components

Boilers have two sections, the combustion section and the heat transfer section.

The **combustion section** is the space in which the fuel-air mixture burns. *Figure 10.1* shows a commercial boiler with the combustion chamber at the bottom. In this boiler, the base is insulated, but the top and sides of the combustion chamber are heat transfer surfaces. The proportion of air significantly influences the efficiency. If there is excess air, it is heated as it goes through the boiler, carrying heat with it up the chimney. Too little air will cause poor combustion, usually producing noxious combustions products and, in the extreme, may cause extra expense by allowing unburnt fuel through the boiler and up the chimney.

The second section of the boiler is the **heat-transfer section**. This section comprises the two upper spaces in *Figure 10.1*, where the hot gases pass right-to-left and then left-to-right, before exiting to go up the flue. In large boilers, the heat transfer section will be fabricated of cast iron sections that are bolted together, or of welded steel plate and tubes. In smaller, particularly domestic, boilers, the heat-transfer section may be fabricated from copper, aluminum or stainless steel sheet. Boilers can be designed for any fuel: electricity, gas, oil, or

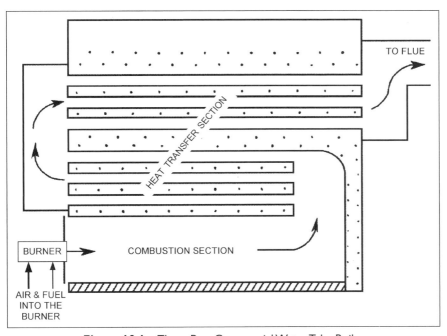

**Figure 10.1** Three-Pass Commercial Water Tube Boiler

138   Fundamentals of HVAC

coal are the most usual. In this age of recycling and sustainability, there is also an initiative to use urban and manufacturing waste as fuel.

In all boilers, there is a need to modulate, or adjust, the heat input. Gas and oil burners may be cycled "on" and "off." The longer the "on" cycle, the greater the heat input. With the "on-off" cycle, the water temperature or steam output will vary up and down, particularly at low loads. This may not matter, but the efficiency improves and cycling effect is much reduced by having a burner with "high-low-off" cycles.

On larger units, a modulating burner will usually be installed that can adjust the output from 100% down to some minimum output. The burner modulation range is called the **"turn-down ratio,"** which is the ratio between full "on" and the lowest continuous operation. A burner that can operate at anywhere from 100% output down to 10% output has a 10:1 turn-down ratio. With a modulating burner, efficiency increases as the output drops. This increase in efficiency is due to the increase in the ratio of heat-exchanger surface-area to heat-input as the output, or **firing rate**, is reduced.

In a coal-fired boiler, the adjustment is achieved by altering the draft of combustion air through the grate. As the air supply increases, the fuel burns faster and hotter, increasing the boiler output.

In general, boiler efficiency drops as the mean temperature of the heated fluid rises. As a result, a hot-water boiler will be more efficient heating water from 150°F to 170°F (mean temperature 160°F) than from 160°F to 180°F (mean temperature 170°F). However, the cooler the mean temperature of the heated fluid, the larger the heat-transfer surfaces must be. Here we have another example of where the designer must consider trading the higher ongoing costs and use of fuel against initial equipment costs.

Because boiler operation is critical for the facility, it is often valuable to have a two boiler system, so that there is always one available for maintenance back up. *Figure 10.2* shows a hot water system with two boilers.

**The boilers**, which are connected in parallel so that one can be valved off and serviced or replaced while the other continues to operate.
**Two pumps**, so that pump failure does not prevent operation.
**A pressure tank** which maintains system pressure and accommodates the changes in water volume as the system is heated up from cold. The

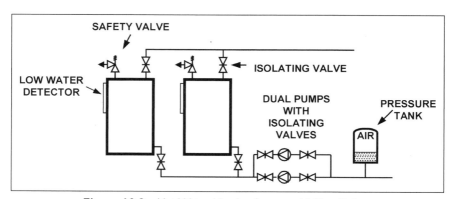

**Figure 10.2**   Hot Water Heating System with Two Boilers

pressure tank often has a membrane in it that separates the water from the air, to prevent absorption of oxygen from the air. If the water level drops too low, more water is pumped into the system; if the pressure needs to be increased, more air is pumped into the top of the tank.
- A **spring-loaded safety valve**, which is provided for each boiler. The valve is set to release at some pre-determined pressure. Then if, for example, the burner controls jammed at full fire, the hot water or steam would be released, protecting the system from bursting.
- A **low water detector/cutout**, which is fitted for each boiler. This safety device prevents the boiler from operating with little, or no, water and overheating, which could easily cause serious damage to the unit.

Dissolved oxygen and other chemicals in normal domestic water can cause severe corrosion and fouling of the heating system, especially with steel pipework. In closed hot water systems, water treatment chemicals may be added as the system is filled. Then, periodically, the system water quality is checked and any needed additional treatment added.

In steam systems, the makeup water must be treated to remove oxygen and dissolved solids before it enters the boiler. This is to prevent the boiler from filling with dissolved solids, since steam (pure water) is continuously boiled off. The steam is very corrosive, so a chemical treatment is included to offset the corrosive characteristics. Thus, there is a need for frequent monitoring, since any failure of treatment can cause problems in the boiler and distribution systems.

With the two boilers in parallel, about half the water will flow through each boiler. If just one boiler is firing, the supply temperature will be based on the average temperature of the return water from the idle boiler and the heated water from the firing boiler. If the supply-temperature requirement equals the temperature that is produced by the operating boiler, then the flow through the idle boiler must be stopped, by closing the inlet valve. For systems with low summer loads, this is ideal since the efficiency is maintained and the idle boiler can be serviced with no interruption of hot-water production.

Note that with steam boilers, if one is running, both will fill with steam to the same pressure. The operating boiler keeps the second boiler hot and ready to fire.

Having considered the heating plant, now let us turn our attention to cooling and consider chillers and cooling towers that, together, provide central chilled water in many buildings.

## 10.4 Chillers

Shown in *Figure 10.3*, is fundamentally the same as the basic refrigeration circuit you were introduced to in *Figure 6.3*, Chapter 6, Section 6.3 except that, instead of the evaporator and condenser being air-cooled, they are now water-cooled.

As you can see in the drawing, there are two flows of water, labeled the chilled water and the condenser water. The water that flows through the evaporator coil gives up heat, and becomes cooler. The cooled water is referred to as "**chilled water**." The water that flows through the condenser, called the "**condenser water**," becomes warmer and is piped away to a cooling tower to be cooled before returning to the condenser to be warmed again.

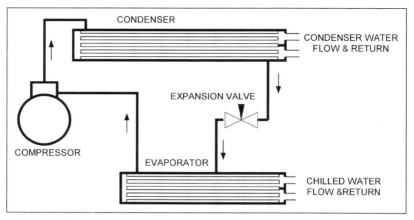

**Figure 10.3** Water Chiller with Water Cooled Condenser

The size of the cooling load determines the requirements for chiller capacity. This requirement can be met by one or more chillers. The standard measure of chiller capacity is the **ton**, a heat absorption capacity of 12,000 Btu per hour. The historical origin of this unit is from the early days of refrigeration, when ice production was the main use. In 24 hours, 12,000 Btu per hour produces one ton (2,000 pounds) of ice. Residential air-conditioners are typically one to three tons; central chillers, delivered as complete, preassembled packages from the factory, can be as large as 2,400 tons; and built-up units can go up to 10,000 tons.

The main difference between chillers is the type of compressor:

Smaller compressors are often **reciprocating units**, very much like an automobile engine, with pistons compressing the refrigerant.

Larger units may have screw or scroll compressors. These compressors are called "**positive-displacement,**" since they have an eccentric scroll or screw that traps a quantity of refrigerant and squeezes it into a much smaller volume as the screw or scroll rotates.

Finally, for 75 tons up to the largest machines, there is the **centrifugal compressor**. It has a set of radial blades spinning at high speed that compress the refrigerant.

The choice of compressors is influenced by efficiency at full and part load, ability to run at excess load, size, and other factors. At times of lower load, the capacity of the reciprocating compressor can be reduced in steps by unloading cylinders. The other types of machine can all have their capacity reduced, to some degree, by using a variable speed drive. In addition, the centrifugal machine has inlet guide vanes that reduce the capacity down to below 50%.

When designing a central plant, it is often worth some additional investment in plant and space to have two 50% capacity chillers instead of a single chiller for the following reasons:

There is 50% capacity available in case of a chiller failing.

The starting current is halved, lowering the demands on the electrical system.

Chiller efficiency is higher, the higher the load on the chiller. When load is lower, the second chiller can be turned off.

Maintenance work can be carried out during the cooling season during times of low load.

A variable chilled water flow arrangement is shown in *Figure 10.4*. The chillers are shown with the condensers dotted, since they are not relevant to the chilled water circuit.

As you can see in the diagram, at full load, both chillers and pumps are running, and the valves in the coil circuits are fully open. As the load decreases, the temperature sensors, in front of each coil, start to close their valve, restricting the flow through the coil. The flow sensor, in the chilled-water pipe from the chillers, senses the flow reduction, and restores flow by opening the bypass valve to maintain chiller flow.

When the load drops below 50%, one of the chillers and pumps can switch off, leaving one pump and one chiller to serve the load. The check valve in front of the pump that is "off" closes, to prevent the chilled water from flowing back through it. The output of each chiller is adjusted to maintain the chilled water set-point temperature. As the cooling load on the two coils drops, the return-water temperature will fall and the chiller will throttle back to avoid over-cooling the chilled water.

Load estimation is quite accurate nowadays, so chillers should be sized to match the estimated load without a 'safety' factor. This is particularly important where there is just one chiller, since it has to handle all load requirements, including low load. If the chiller is a little undersized, there will be a few hours more a year when the chilled water temperature will drift up a bit. This is generally far better than over-sizing. Over-sizing costs more in chiller purchase price, larger pumps, and other components. The larger chiller will have a lower operating efficiency, so it will have a higher operating and maintenance cost, as well as more difficulty dealing with low loads.

If failure to meet the load is critical, such as in some manufacturing operations, then the issue of sizing to the load is combined with the issue of having standby

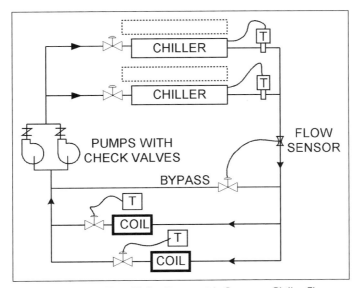

**Figure 10.4** Two Chiller Piping with Constant Chiller Flow

capacity for a failed machine. In this case the manufacturing operation should have two units sized to 50% of the load each, with a third 50% unit as standby.

## 10.5 Cooling Towers

Cooling towers are a particular type of big evaporative cooler.

The following description details the sequence of activity in the natural-draft tower, shown in *Figure 10.5*:

1. Hot water (typically at 95°F,) is sprayed down onto an extended surface "**fill**." The fill normally consists of an array of indented plastic sheets, wood boards, or other material with a large surface area.
2. The water coats the fill surface and flows down to drop into the sump at the bottom.
3. At the same time, air is entering near the bottom and rising through the wet fill.
4. Some of the descending water evaporates into the rising air and the almost saturated air rises out of the tower.
5. The latent heat of evaporation, absorbed by the water that does evaporate, cools the remaining water.
6. The cooled water in the sump is then pumped back to the chiller to be reheated.

The cooling performance and consistency of operation under various weather conditions can be greatly improved by using a fan to either drive (forced draft) or draw (induced draft) the air through the cooling tower. The addition of a fan increases the speed of the air flowing through the tower, and smaller water drops

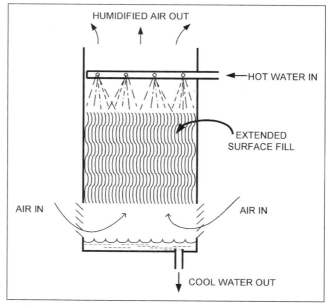

**Figure 10.5** Typical Natural-Draft Open Cooling Tower

may become entrained in the air stream. These drops, if allowed to escape, would be wasted water and could cause wetting of nearby buildings or vehicles. Therefore an array of sheets, called "**drift eliminators**" is included to catch the drops and return the water to the spray area.

In the open cooling tower, the condenser water is exposed, or open, to the air and it will collect dirt from the atmosphere. Strainers will remove the larger particles but some contamination is inevitable. This contamination can be avoided by using a closed-cooling tower, as is shown in *Figure 10.6*. Here, the fluid to be cooled is contained in a coil of pipe in place of the fill. This closed tower is an induced-draft tower (the fan draws the air through the tower) and includes drift eliminators.

The figure shows water in the closed coil. Alternatively, refrigerant can be passed through the coil and then the refrigerant pipe loop in the tower is the refrigerant circuit condenser.

In a typical cooling tower, at full load, the closed circuit fluid, water or refrigerant, can be cooled 30–35°F cooler than with an air-cooled coil. This substantially increases the performance of the refrigeration system.

Now that you understand the physical arrangement of the cooling tower, let us consider what is going on inside of the tower. *Figure 10.7* shows the basic operation of the cooling tower. On the left, the warm water is falling and becoming cooler while on the right, air rises through the tower and becomes more saturated with water vapor. The evaporating water absorbs its latent heat of evaporation from the surrounding air and water before it is carried up and out of the tower in the flow of air. In effect, the air is a vehicle for removing the evaporated water.

The cooling performance of the tower is dependent on the enthalpy of the ambient air entering the tower. Remember, the drier and cooler the air,

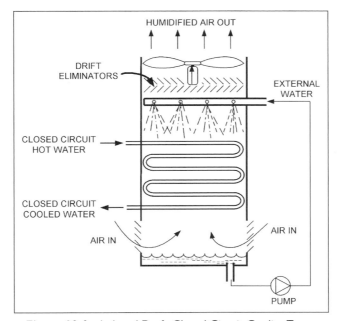

**Figure 10.6** Induced Draft, Closed Circuit Cooling Tower

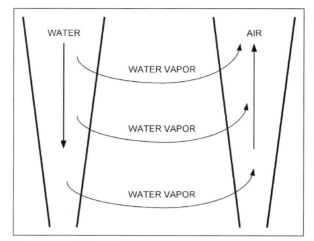

**Figure 10.7** Flow of Water, Water Vapor, and Air in a Cooling Tower

the lower its enthalpy. The lower the enthalpy of the entering air, the greater the evaporation, and therefore, the greater cooling performance.

Surprisingly, the temperature of the air may rise, stay the same or fall as it passes upwards through the tower.

Look at *Figure 10.8*, and consider these **two scenarios**:

**Scenario 1**: Air at Condition 1, enters the tower and is heated and humidified as it rises through the tower, to leave the tower virtually saturated at Condition 3. As the water cools, it provides heat to raise the air temperature.

In this first situation, from Condition 1 to Condition 3, the amount of water evaporated to absorb latent heat was equal to the reduction in the water enthalpy less the cooling provided by the cool air being warmed:

*Total latent heat of evaporation = Reduction in water enthalpy − air cooling effect*

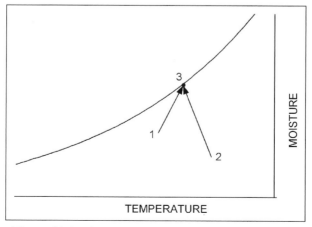

**Figure 10.8** Cooling Tower Psychrometric Chart for Air

**Scenario 2**: In contrast, when warmer air, at roughly the same enthalpy, enters the tower at Condition 2, it will be cooled and humidified as it passes through the tower to leave at Condition 3. The reduction in air temperature is achieved through additional evaporation.

In this situation, from Condition 2 to Condition 3, the amount of water evaporated to absorb latent heat was equal to the reduction in enthalpy of the water plus the heat required to lower the air temperature:

*Total latent heat of evaporation = Reduction in water enthalpy + air heating effect*

Overall, the tower has approximately the same cooling effect on the water for entering air with the same enthalpy whatever the entering air temperature. However, with the same enthalpy, as the air becomes hotter and dryer more evaporation will take place.

The tower capacity can be reduced in several ways. The fan can be cycled on-and-off, but the frequent starts are very hard on the motor. Better, for both energy conservation and motor life, is to use a two-speed motor and cycle between high, low and off. For slightly better control and energy savings, a variable speed fan can be used.

The water that is evaporated leaves behind any dissolved chemicals. At full load this can be as much as 1% of flow. In addition, the water cleans the air, removing dust and debris. Since the water is warm and full of nutrients, it is an ideal site for bacterial growth, legionella in particular. It is thus critical that the tower is regularly cleaned of dirt buildup and treated, to prevent biological growth.

## The Next Step

We have considered components, systems and, in this chapter, central plant. Along the way, equipment has been 'controlled' and energy saving has been mentioned. In the next chapter, Chapter 11, we will focus on controls and how they work. We will revisit several of the systems you have already learnt about, and consider their particular control features. Then after controls, we will consider energy conservation in Chapter 12.

## Summary

This chapter has been concerned with central plant, specifically with boilers, producing steam or hot water, chillers producing chilled water and cooling towers that cool the chillers.

### 10.1 Introduction

Central plants generally require less maintenance than numerous smaller package systems and the equipment usually has a longer life. Other advantages include ease of operation and maintenance in a central location; efficiency; heat recovery options; less maintenance and a longer life. Cons include: cost of installation, space requirements for the equipment and for the distribution pipes. Issues of seasonal efficiency were also raised.

## 10.2 Central Plant Versus Local Plant in a Building

Issues that can influence the choice include installation costs vs. operating costs. For minimum installation cost, the package approach usually wins. However, the central plant has several operational benefits.

## 10.3 Boilers

Boilers are pressure vessels used to produce steam or hot water. The critical design factor for boilers is pressure. A low-pressure steam boiler operates at a pressure of no more than 15 psig. Low-pressure hot water boilers are allowed up to 160 psig.

Boilers and system components are covered by local code requirements. The safety equipment and staff monitoring requirements are far less stringent for low-pressure boilers so there is a significant incentive to use low-pressure.

Boilers have two sections: The combustion section is the space where the fuel-air mixture burns; the second section of the boiler is the heat transfer section. In all boilers there is a need to modulate the heat input. On smaller units, the efficiency improves and cycling effect is reduced by having a "high-low-off" burner. On larger units, a modulating burner can adjust the output from 100% down to some minimum output. The burner modulation range is called the "turn-down ratio." With a modulating burner, efficiency increases as the output drops and efficiency drops as the mean temperature of the heated fluid rises.

Boilers can run in parallel: With two water boilers, about half the water will flow through each boiler; with steam boilers, if one is running both will fill with steam to the same pressure.

In steam systems, there is a constant loss of water in the condensate return system. To prevent problems with solids build-up in the boiler and distribution pipe corrosion, continuous high quality water treatment is required.

## 10.4 Chillers

Chillers are refrigeration machines with water, or brine, heating the evaporator. The standard measure of chiller capacity is the ton, a heat absorption capacity of 12,000 Btu per hour. The main difference between chillers is the type of compressor. Smaller compressors are often reciprocating units, larger units may have screw or scroll positive-displacement compressors, and for 75 tons up to the largest machines, there is the centrifugal compressor.

Chillers should be sized to match the estimated load without a 'safety' factor. An oversized chiller will have a lower operating efficiency, so it will have a higher operating and maintenance cost, as well as more difficulty dealing with low loads. When designing a central plant, it is often worth having two 50% capacity chillers instead of a single chiller. If failure to meet the load is mission critical, use two units sized to 50% of the load each, with a third 50% unit as standby.

## 10.5 Cooling Towers

Cooling towers are a particular type of big evaporative cooler. In the cooling tower, warm water is exposed to a flow of air, causing evaporation and therefore, cooling of the water.

The psychrometric chart can be used to illustrate the workings of the cooling tower.

It is often considered worthwhile to over size the tower to ensure that full chiller capacity will always be available. The tower capacity can be reduced: by using a fan that can be cycled on and off; with a two-speed motor that can cycle between high, low and off; or a variable speed fan can be used.

A danger of cooling towers arises from the warm, nutrient rich environment that can propagate bacteria growth, therefore, the tower should be regularly cleaned of dirt buildup and treated, to prevent biological growth. In addition, some water must be bled off to prevent the build-up of dissolved solids.

## Bibliography

1. 2004 ASHRAE Systems and Equipment
2. 2003 ASHRAE HVAC Applications
3. (ASME) Boiler and Pressure Vessel Code

Chapter 11

# Controls

## Contents of Chapter 11

Study Objectives of Chapter 11
11.1 Introduction
11.2 Basic Control
11.3 Typical Control Loops
11.4 Introduction to Direct Digital Control, DDC
11.5 Direct Digital Control of an Air-Handler
11.6 Architecture and Advantages of Direct Digital Controls
The Next Step
Summary
Bibliography

## Study Objectives of Chapter 11

Chapter 11 starts off by describing the basics of control and introducing you to some of the terminology of HVAC controls. After this introduction, we consider the physical structure and software of Direct Digital Control, DDC, systems. In this section, we demonstrate some of the control possibilities that are available with DDC by revisiting some of the references to controls in earlier chapters. Finally, there is a brief introduction to the architecture of DDC systems and their advantages. After studying the chapter, you should be able to:

Explain the following terms: normally open valve, modulating, proportional control, controlled variable, setpoint, sensor, controller, and controlled device.
Describe an open control loop and a closed control loop and explain the difference between them.
Explain how the DDC system replaces conventional controllers.
List the four main DDC point types and give an example of each one.
Explain how the knowledge in a DDC system can be put to good use.

## 11.1 Introduction

Every piece of equipment that we have introduced in this course requires controls for operation. Some equipment, such as a rooftop package unit, will likely come with factory-installed controls, except for the thermostat. The thermostat has to be mounted in the space and wired to the packaged unit. In other built-up systems, every control component may be specified by the designer and purchased and installed under a separate contract from the rest of the equipment.

Whether the controls are a factory package or built-up on site, well-designed controls are a critical part of any HVAC system. The controls for a system may differ from project to project for a number of reasons. Design considerations for controls choices include availability of expertise in maintenance and operations of the controls, repair and maintenance expense budgets and capital costs of control equipment.

To elaborate, one should always choose controls that are suited to the available maintenance and repair expertise and availability. Find out how the client will be arranging maintenance of the system. As an example, it is generally unwise to choose the latest and greatest high-tech controls for a remote school, unless the school has a maintenance system in place to support the controls. It is generally better to aim for simplicity and reliability in this type of situation.

On the other hand, if the client has experienced, well-trained, controls staff available, on site or by contract, there is an opportunity to specify something quite sophisticated. As always, economics plays a controlling role and the challenge is to demonstrate how the sophisticated computerized system will perform better and save energy compared to a simple off-the-shelf option.

There are several types of controls and each has specific features that make it by far the best choice in particular circumstances. The following is a brief introduction to the main types.

## Control Types

Controls fall into broad categories based on a particular feature.

**Self-powered Controls** require no external power. Various radiator valves and ceiling VAV diffusers have self-powered temperature controls. These units are operated by the expansion and contraction of a bellows that is filled with a wax with a high coefficient of expansion. As the temperature rises, the wax expands, lengthening the bellows. This closes the radiator valve (cuts back on heating) or opens the VAV diffuser (increases the cooling). The advantage of these units is that they require no wiring or other connection so installation cost is minimal.

**Electric Controls** are powered by electricity. We will introduce two types of electric controls in this course:

*On/off Electric Controls* are used in almost every system to turn electrical equipment on and off. The electric thermostat is the most common example.

*Modulating Electric Controls* are based on small electric motors and resistors that provide variable control.

**Pneumatic Controls** are controls that use air pressure: the signal transmission is by air pressure variation and control effort is through air pressure on a diaphragm or piston. For example, a temperature sensor may vary the pressure to the controller in the range of 3 psig to 15 psig (pounds per square inch gauge). The controller will compare the thermostat line pressure with the setpoint pressure and, based on the difference, adjust the pressure to the heating valve to open, or to close, the valve. The heating valve will typically have a spring to drive it fully open and the increasing air pressure will close the valve against the spring. The valve is called a "**normally open**" valve, since failure of the air system would have air pressure fall to zero and the spring would open the valve. A "**normally closed**" valve is the opposite, with the spring holding it closed until the air pressure opens the valve.

Pneumatic controls require a continuous source of compressed air at 15 psig for sensing and controlling. When considering the total cost of the pneumatic system, the provision of the compressor(s), the operation and maintenance cost, and the energy lost with leaks have to be factored into the total cost. However, the pneumatic system does have the advantage of relatively inexpensive and powerful actuators (a device that moves a valve or damper) and it is relatively easy to learn to maintain and service.

**Electronic Controls**, or more correctly **Analogue Electronic Controls**, use varying voltages and currents in semiconductors to provide modulating controls. They have never found great acceptance in the HVAC industry, since Direct Digital Controls offered much more usability at a much lower price.

**Direct Digital Controls, DDC**, are controls operated by one, or more, small computer processors. The computer processor uses a software program of instructions to make decisions based on the available input information. The processor operates only with digital signals and has a variety of built-in interface components so that it can receive information and output control signals.

There are many instances where the types of controls are mixed. For example a DDC system could have all electric "**sensors**," the units that measure temperature, humidity, pressure or other variable properties. This same system may also have pneumatic actuators on all the valves, since pneumatics provide considerable power and control at low cost. A "**transducer**" creates the interface between the electrical output of the DDC system and the valve. The transducer takes in the DDC signal, say a voltage between zero and ten volts, and converts it to an output of 3psi to 15psi. Thus, at zero volts the output will be 3psi, rising to 15psi at ten volts.

We will spend considerable time on DDC controls later in the chapter. For now, let us consider the basics of controls—what makes them work.

## 11.2 Basic Control

You instinctively know about control. You control all sorts of actions in your daily life. In this section we are going to introduce the basic ideas of controls. Your understanding of the rest of the chapter depends on you being really comfortable with the ideas in this section. Take the time to think about the ideas presented and how controls operate.

We are going to start with the simplest of controls, "on-off." As the name implies, the element being controlled is either "on," or "off."

> Consider a domestic hot water tank with a thermostatically controlled electric heating element near the bottom. Water becomes less dense as it is heated above 39.4°F, so, as the element heats the water, hot water will rise to the top. When the water at the thermostatically controlled element is hot, all the water above it is hot and the thermostat will turn off the heating element.
>
> Now, let us assume someone runs a little hot water. Cold water enters the bottom of the tank and cools the thermostat. The thermostat switches the element "on" and soon the tank is filled with all hot water once more. Suppose that later, one person runs a shower

as someone else is running hot water for washing clothes. Although the element will come on, it can't keep up with this large load. Very soon all the hot water is gone and the tank is full of cold water. Both users turn off the taps. The element at the bottom of the tank will slowly heat the whole tank back up to the required temperature.

In order to achieve a quick recovery of hot water, we need a second element near the top of the tank. An element near the top of the tank only has to heat the water above it, so it will get a small amount of water up to temperature much more quickly. Now we have two elements. Do we need them on at the same time? No. If the top element is needed, the bottom element is not needed. So, when the top element turns on, for quick recovery, it also breaks the circuit to the bottom element. Once the water above the top element is hot, that element switches "off" and the bottom element switches "on" to heat the rest of the water in the tank. This give us an 'either/or' control decision – either the top or the bottom element can be "on."

The result is a tank that heats a little water quickly, and, in a much longer time, heats the full tank, with the electrical load of one heating element.

This is a simple example of how "on-off" controls can be cascaded to produce simple, but very effective, control. With some ingenuity, quite complex and extremely reliable electric controls can be developed.

Now let us move on from "on-off" to "**modulating**" controls. Modulating means 'variable'. One type of modulating control is **proportional control**. This is best explained with a 'hands-on' demonstration.

> Take a jug and fill it with water.
> Take a tumbler, and place it in a spot that won't be damaged if water overflows (like in a sink).
> Next, see how quickly you can fill the glass so that the water level is right up at the rim of the glass–so full you'd need a very steady hand to drink it!

If you didn't actually do the task, take a few moments to relax, and visualize the empty glass with the full jug on the table beside it. The jug is heavy as you pick it up and start to pour. You hear the water flowing in, feel yourself tipping the jug, see the level rising and feel that tension as you slow the flow to drips, to make it just reaches the top, and then you stop pouring.

What happens? When you see the glass empty or just starting to fill, the water level is a long way from the rim. Naturally, you start pouring quite quickly. As the glass fills, you slow the flow until you're just *dripping* the water in, to get the glass quite full, to the rim, without going over. The change in rate at which you pour is roughly proportional to the distance of the water from the rim. This change in rate is called the **gain**. You are acting as a "**proportional controller.**"

Proportional control is the basis of the majority of control loops—the rate is proportional to the distance from the target—the setpoint.

Now imagine this more complicated scenario of proportional control. In this scenario, we will be demonstrating **offset** and **overshoot**.

> Someone has attached a hose to the bottom of your glass and runs it to a tap downstairs, out of your sight. Your job, now, is to keep the glass full. You fill the glass, and then you notice the level is

dropping slowly. In response, you start to pour slowly, just keeping the glass near full. Then, you realize the level is dropping faster, so you tip the jug. Suddenly the glass is full–it is overflowing!

What happened?

Initially, the hose tap was opened just a little, and it was easy for you to pour slowly to keep the level near the rim. When the tap was opened wide, though, the jug had to be tipped a lot to keep up, so when the tap was suddenly closed, it took a moment for you to realize that the water level in the glass was rising rapidly. It took another moment—too long—to straighten the jug and stop the flow, and the water overflowed over the top of the glass.

Just like you, a control system has an easy time with slow steady changes. Note though, you had to notice that the water level had dropped before you started to pour. This created a time delay. Note also that you attempted to keep the glass almost full rather than totally full. This represented an **"offset"** from target, the setpoint. Then, when the glass started to drain rapidly you poured faster, to keep it from being empty, rather than trying to maintain the level just at the rim – even more offset.

Finally, the drain on your glass stopped, and you were too slow to straighten the jug and stop the flow. The water overflowed – serious **"overshoot"**! This overshoot could have been reduced if you had been restricted on how fast you could pour. If you had less **gain** you would have had less ability to keep up with sudden changes and the overshoot would have been much less.

You now have some feel for controls and what they do. There are many added refinements to controller action that are explained in the ASHRAE Course, *Fundamentals of Controls*[1].

Now lets consider some real HVAC examples.

There are two types of control **"closed loop"** and **"open loop."** Let us start by considering the main components of a closed loop control as shown in *Figure 11.1*.

The top half of the figure illustrates a simple air heating control loop. A temperature sensor measures the temperature of the heated air and sends that information to the controller. The controller is also provided with the required setpoint (similar to the setting on the front of a room thermostat). The controller first compares the measured temperature with the setpoint and, based on the difference, if any, generates an output signal to the valve. If the sensed temperature were a little higher than the setpoint, the controller would generate an output to close the valve a little. The valve would close, reducing the heating coil output. The air would be warmed less and the temperature sensor would register a lower temperature and sends that information to the controller—and so on round and round the closed control loop.

The lower part of the figure is the same process with the generic names for the parts of the control loop.

- The **"controlled variable"** is the variable, in this case, temperature, that is being controlled. Controlled variables are typically temperature, humidity, pressure and fan or pump speed.
- The **"setpoint"** is the desired value of the controlled variable. In this example it is the air temperature that is required.

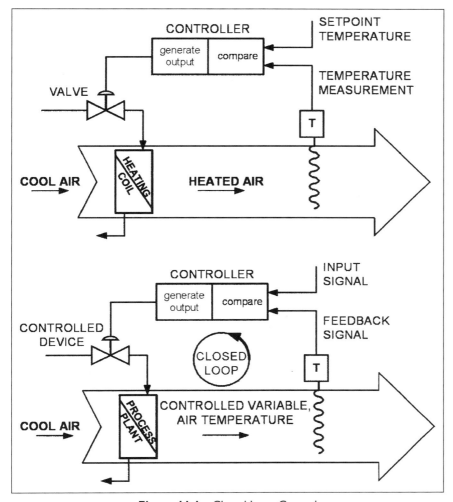

**Figure 11.1** Closed Loop Control

The "**sensor**" measures the controlled variable and conveys values to the controller. In this case the sensor measures temperature.

The "**controller**" seeks to maintain the setpoint. The controller compares the value from the sensor with the setpoint and, based on the difference, generates a signal to the controlled device for corrective action.

Note that a room thermostat contains the temperature setpoint, which is your adjustment of the setting on the front of a room thermostat. It also contains the room temperature sensor and the controller. A humidistat is the same, except that it is sensing relative humidity.

The "**controlled device**" responds to signals received from the controller to vary the process—heating in this example. It may be a valve, damper, electric relay or a motor driving a pump or a fan. In the example it is the valve controlling hot water or steam to the coil.

154    Fundamentals of HVAC

To make sure you understand the above definitions, think about the simple example of pouring water in to fill the glass. What do you think were the following?

Setpoint
Sensor
Controller
Controlled device
Controlled variable

Answers are included at the end of this section.

So far, we have been discussing closed loop control—based on feedback, the controller makes continuous adjustments in order to maintain conditions that are close to the setpoint.

Another type of control called "**open loop**" control, where there is no feedback. Consider a simple time clock that controls a piece of equipment. The time clock is set to switch "on" at a specific time and switch "off" at a later time. The time clock goes on switching "on" and "off" whether the equipment starts or not. In fact, it will go on switching even if the equipment is disconnected. There is no feedback to the time clock, it just does what it was set to do.

In Chapter 8 we introduced the idea of "outdoor reset." Outdoor reset is a method of adjusting the temperature of a heating source, or cooling source, according to changes in outdoor temperature. This is an example of open loop control.

We are going to add outdoor reset to our air heating system, as illustrated in *Figure 11.2*, below.

*Figure 11.2* illustrates the same closed control loop as in *Figure 11.1*, but with outdoor reset added. The ambient (outdoor) temperature sensor provides

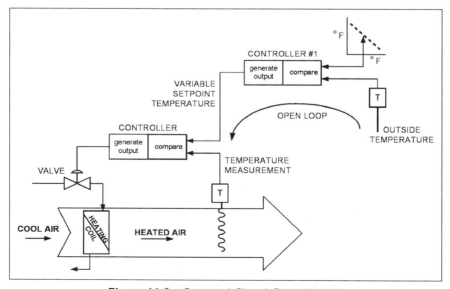

**Figure 11.2**   Open and Closed Control Loops

Controller #1 with a signal, and the setpoint is provided as a variable according to the outdoor temperature. This is illustrated as a little graph, in the top right hand corner, showing a falling supply setpoint temperature (Y axis) as the temperature rises (X axis). The output of Controller #1 is the setpoint for our closed loop controller. The open loop measures temperature and provides an output. It has no involvement with the result; it just does its routine – open loop control – no feedback.

Alternatively, we could have chosen to use a chilled water system and to use outdoor reset to raise the chilled water temperature as the outside temperature dropped.

Outdoor reset is a common requirement, so manufacturers frequently package the two controllers into one housing and call it a '**reset controller**'. This packaging of several components of the control loop is similar to the thermostat package where the setpoint, the temperature sensor and the controller are packaged in one little box.

Answers: Setpoint: rim of glass; Sensor: eyes; Controller: you; Controlled device: jug; Controlled variable: water depth.

## 11.3 Typical Control loops

Having considered the basics of control loops in the previous section, now lets look at some real, more complete, control loops. We will start by adding time control, another open loop, to our previous example, as *Figure 11.3*.

A time clock now provides power to the controllers according to a schedule. Typical commercial thermostats include the 5–1–1 time clock function. 5–1–1 means that they have independent time schedules for the 5 weekdays, 1 for Saturday, and 1 for Sunday.

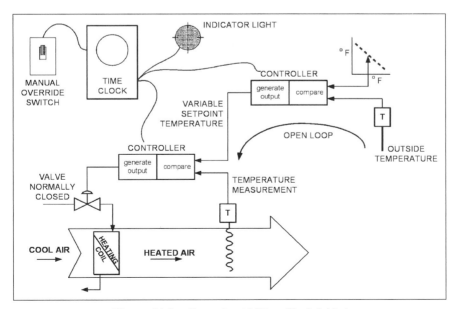

**Figure 11.3** Controls with Time Clock Added

156   Fundamentals of HVAC

The system shown also has a manual-override switch that allows the occupant to switch the system 'on' when the time clock has it 'off'. There is an obvious energy waste issue here, since the occupant may forget to switch back to the time clock. In most time clocks, the manual switch is part of the unit, rather than a remote switch as shown in the figure.

In addition, in the figure, there is an indicator light to show that the system is 'on'. When the time clock switches 'on', it provides power to the lamp and power to the controllers. It has no idea whether the controllers are 'on' or even whether they are connected! The lamp does not indicate that the system is working. What it indicates is that power from the time clock is available. This type of open-loop indication is very common. If you are involved in trouble shooting equipment, think about the real information provided. Even if the lamp 'off ', it does not mean there is no power to the controllers–the lamp could have burnt out!

In our diagram there is just one heating coil being controlled. In many packaged units, there will be two stages of cooling and two stages of heating. A single 5–1–1 thermostat will provide full control, turning 'on' one stage of cooling, and then the second stage, **or** one, and then the second stage of heating. The really good feature of a single, packaged thermostat is that there cannot be any overlap of control. For example, if a separate thermostat were used for heating and another for cooling, they could mistakenly be set so that the first stage of heating was 'on' when the first stage of cooling starts – a real waste of energy.

This issue of staging controls so that energy use is minimized is important in many areas. An example is the sequencing of control in a VAV box with a reheat coil. A VAV system provides cold air for cooling and ventilation. Should a zone require less cooling than is provided at the minimum airflow for ventilation, then the reheat coil is turned on. In the control system for the box, there are two important requirements:

1. The heating coil must only be activated at minimum airflow.
2. There must be minimal cycling between 'coil on' and 'coil off'.

To achieve this, a single controller is used to control both the airflow and the coil, in sequence. The heating valve and volume damper are normally closed. The volume damper has a minimum setting for minimum ventilation.

As an example, the box and controller actions are shown in *Figure 11.4*. Starting on the left, when the space is cold, the controller opens the heating valve fully. As the zone warms up, the controller closes the heating valve.

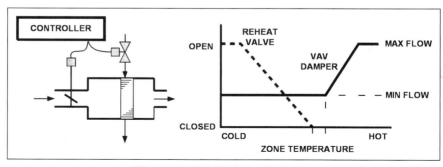

**Figure 11.4**   VAV Box with Reheat

Once the heating valve is closed, there is a dead band of temperature change (no heating, no additional cooling) before the controller starts to open the volume damper to increase the cooling up to maximum.

In addition to the simple control loops we have discussed, there are more complex loops that have many inputs. Staying with VAV for a moment, there are many systems where the fan speed is controlled by the requirements of the VAV boxes. A system, for example, might have 50 or more boxes. We want each to have enough air but we don't want to run the fan any more than needed. To manage this, we need to know when each box has adequate air flowing through it.

A VAV box has enough air if its damper is not fully open. Thus, we would be very confident that if every box has its damper at less that 95% open, there is enough air pressure in the system. However, determining if every box meets this condition is only practical in a DDC system. We will begin to examine these DDC systems in the next section.

## 11.4 Introduction to Direct Digital Control, DDC

As briefly mentioned earlier, small computer processors operate Direct Digital Controls, DDC. 'Digital' means that they operate on a series of pulses, as does the typical PC computer. In the DDC system, all the inputs and outputs remain, however, they are not processed in the controllers, but are carried out in a computer, based on instructions called the "**control logic**."

Figure 11.5, which follows, is the same control diagram that we saw in *Figure 11.3*, but with the controlling components, (the time clock, and the two controllers), blanked out. All the system that has been blanked out is now replaced by software activity in the computer.

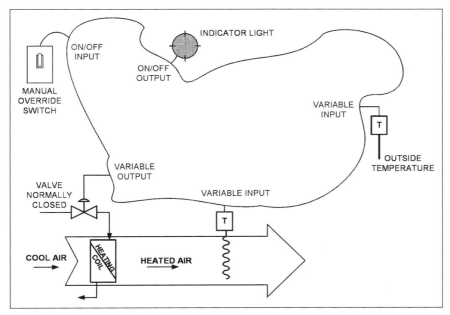

**Figure 11.5** Control Scheme (from 11–3) without Controlling Components

158    Fundamentals of HVAC

In *Figure 11.5*, each input to or output from the DDC computer has been identified as one of the following

On/off input – **manual switch**
On/off output – **power to light**
Variable input – **temperature from sensor**
Variable output – **power to the valve**

These are the four main types of input and output in a control system. Lets consider each one briefly in terms of a DDC system.

**On/off input.** A switch, a relay, or another device closes, making a circuit complete. This on/off behavior has traditionally been called "digital." Therefore in DDC terms it is generally called a "**Digital Input**," or **DI**.

The term 'digital' is not considered technically correct, since there is no series of pulses, just one 'on' or 'off'. Thus, for on/off points the term "Binary" is considered more correct, and the term is being encouraged in place of 'digital'. So, "Binary Input," BI, is the officially approved designation of an 'on/off' input.

**On/off output.** The on/off output either provides power or it does not. The lamp is either powered, 'on', or not powered, 'off'. In a similar way, this is called a "**Digital Output**," **DO** or more correctly, binary output, **BO**.

**Variable input.** A varying signal, such as temperature, humidity or pressure, is called an "analogue" signal. In DDC terms, the input signal from an analogue, or varying, signal is called an "**Analogue Input**," or **AI**.

**Variable output.** In the same way, the variable output to open or close a valve, to adjust a damper, or to change fan speeds, is an "**Analogue Output**," **AO**.

You might think the next step is to connect these DI, DO, AI, AO points to the computer. Things are not quite that simple. A sensor that measures temperature, produces an analogue, varying signal and our computer needs a digital signal. So between each AI device and the processor there is an "**A/D**," "**analogue to digital**," device. These A/D devices, for AIs, are usually built in with the computer.

Similarly, for AO points there is a "**D/A**," or "**digital to analogue**," device that converts the digital signal to a 0–10 volts or 4–20 milliamp electrical signal. This signal has too little power to operate a valve or damper. If, for example, the controlled device is a valve that is powered by compressed air, the analogue electrical signal will go to a "**transducer**" in which the electrical signal will be converted to an air pressure that drives the valve. If the valve is electrically powered, the transducer will convert the low power, analogue signal to a powerful electric current.

Only standard telephone cable is required to carry the analogue electrical signal, hence the transducer is often separate from the processor and close to the controlled device. This is because it is far less effort, and cost, to run standard telephone cabling to the transducer rather than to run the air line (or electric power cable) to the processor location and back to the valve.

## Naming Conventions

In a DDC system every input and every output must have a unique name. There are a variety of naming conventions depending on personal preference and the size and complexity of the system. Many are based on a hierarchy of elements such as

*Type – Building – System – Point – Detail*

'Detail' allows for a number of identical points, VAV boxes for example. If we assume our build is called 'NEW', our points list might be:

| | |
|---|---|
| AI NEW AH1 OAT | AI, in NEW, on air-handler1, outside air temperature |
| AI NEW AH1 DT | AI, in NEW, on air-handler 1, duct temperature |
| AO NEW AH1 DT | AO, in NEW, on air-handler1, duct temperature control |
| DI NEW AH1 MAN | DI, in NEW, on air-handler1, manual control |
| DO NEW AH1 IND | DO, in NEW, on air-handler1, indicator light |

## Sequence of Operations

Now look back at *Figure 11.5*. As we noticed earlier, the controllers and time clock are all blanked out. In a DDC system, all the actions of the controllers and time clock are carried out through software in the small computer processor. The software is a set of ordered operations, which is often called the "**sequence of operation**." What do we require our software to do? The following is a very simple 'English Language' sequence of operations.

Do the following things:
If the time is between xx:xx a.m. and yy:yy p.m. run mode is 'ON', otherwise run mode is 'OFF'
If the manual switch is closed, DI, run mode is 'ON'
If run mode is 'OFF' close heating valve
If the system is in run mode 'ON', do the following commands
Check ambient temperature, AI, and remember the value as 'ambient'
Using 'ambient' lookup required setpoint from (graphic) schedule to find required air setpoint temperature. Remember this value as 'setpoint'
Check air temperature in the duct, AI, and remember it as 'temperature'
If 'temperature' is less than 'setpoint' increase output to valve, AO
If 'temperature' is greater than 'setpoint' decrease output to valve, AO
Go back to the beginning

These instructions are typically written into the DDC processor using a standard personal computer, PC. The programming may be a more formal version of our little example, or may use graphic symbols instead. The DDC processor can also be programmed to sound alarms, issue warnings, write messages, plot graphs, and draw graphics through the PC.

So now we can redraw *Figure 11.5*, to show the DDC system, *Figure 11.6*.

As it is shown, there is no way of accessing the processor. In a real system, there is a communication connection providing access from a computer, typically a desktop PC or a laptop computer. The PC has many names including "the operator interface," "front end," or "operator machine interface" (OMI). Assuming that this panel has everything it needs to run the system, it is called a "**standalone panel**." Standalone means it has everything to keep running on its own.

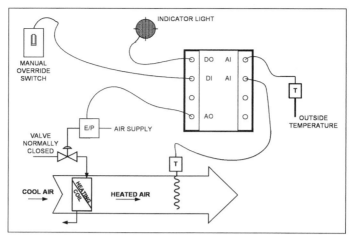

**Figure 11.6** DDC Control Schematic

A really important thing to understand is that the DDC controller can record what happens over time and either directly use that information in useful ways or provide it to the operator.

In our simple system, for example, the DDC system could check how many hours the manual switch had been on. If it had been on more than three hours, it could issue an alarm to the PC, asking the operator if it should still be in manual. This alarm could repeat every two hours to remind the operator to change back to the schedule.

In addition, there are some faults it could be programmed to detect. The heating valve is normally closed when the system is 'OFF'. When the system starts, the duct temperature should be no hotter than the building or the outside ambient temperature. Now, our system does not know the building temperature, but we could assume it would be no higher than 80°F. Thus, we could have a software routine that checked, on startup, that the duct temperature was both no higher than 5°F above ambient, and no higher than 85°F. If the duct temperature were above both these two checks, it could issue a warning that the heating valve may not be shutting off completely.

It is this ability to collect information about every point and to process it, that makes DDC so powerful. Treating it as only a controller replacement is to miss out on the real power of the system.

Lets consider a very simple illustration of this power of knowledge that can be written into a DDC system. We are going to consider two offices served by a single VAV box as illustrated in *Figure 11.7*.

The objective is to provide the occupants with conditions that are as comfortable as possible. If we connect an occupancy sensor and temperature sensor in each office, the DDC system will know if the office is occupied and the current temperature in each office. When both offices are occupied, the system can average the temperatures of the two offices and keep the average as close to setpoint as possible. Now, when the occupancy sensor detects that one of the offices is vacated, the controller can wait a few minutes to avoid annoyance and then slowly change to controlling based on just the occupied office temperature.

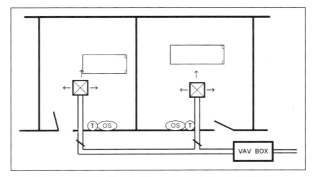

**Figure 11.7** Two Offices Served by one VAV Box

In addition to improving the temperature control, the occupancy sensors also allow the system to modify the amount of outside air being brought in. If one office is vacated, the outside-air volume can be reduced by the assigned volume for the empty office.

Finally, when both offices are empty, the system does not need to maintain the temperature to the same tight limits, and there is no requirement for ventilation air, so, if there is no thermal load, the VAV box can be completely closed. Similar to the example of $CO_2$ control, in Chapter 4, Section 4.5.1, the system only provides service to occupants who are present.

There is one more advantage. The system can be designed to prevent the lights being left on for long periods when the office is unoccupied. One method is to provide power to the lighting circuits (not switch them on, just provide power) when the room is sensed as being occupied. The occupant can switch the light on and off when they like, but when they leave, it will soon go out. The system delays turning the light off for several minutes, to avoid annoyance when the occupant is only away for a few minutes.

This section has introduced you to basic ideas of DDC.

- The sensors and actuators stay, but all the control logic is in the software.
- There are four types of input and output, DO (BO), DI (BI), AO and AI.
- The complete software is a set of instructions that the DDC system can interpret and act upon.

In the next section, we are going to consider the points and sequence of operation of an air-handler. Then in Section 11.6 we will consider how DDC units can be interconnected, and can share information with each other and the operators to make a full-scale control system, rather than a collection of control loops.

## 11.5 Direct Digital Control of an Air-Handler

In this section we are going to consider a constant-volume air-handler serving a single zone, designated '001'. The air handler uses space temperature for control, with no mixed air control, unlike air-handlers that we have discussed before. This is a design choice, unless there is a local code that requires a specific method. Where ASHRAE/IESNA Standard 90.1-2004 *Energy Standard for*

162    Fundamentals of HVAC

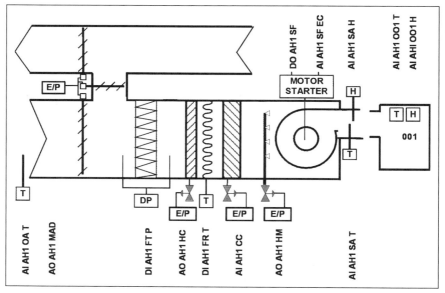

**Figure 11.8**   System Schematic

*Buildings Except Low-Rise Residential Buildings*[3] is incorporated into the local building code requirements, the use of mixed air control is not allowed. We will discus this standard in the next chapter.

To specify a DDC control system, ideally, one produces three things:

1. A schematic of the system with the control points labeled, *Figure 11.8*
2. A list of control points with their characteristics, *Figure 11.9*
3. A schedule of operations

The schematic with the points labeled is not always provided, but it can avoid arguments about the location of points after installation, and it provides the maintenance staff with a map for locating points.

### Sequence of Operation

**Schedule**: Provide calendar/time schedule with minimum of three occupied periods each day.

**Unoccupied**: When calendar schedule is in unoccupied mode, and if space temperature is above 60°F, the fan shall be off, heating valve closed, cooling valve closed. If space temperature falls below 60°F, then the outside dampers and cooling valve to stay closed, heating valve to 100% open, and start fan. When space temperature reaches 65°F, turn fan off and heating valve closed.

**Occupied**: When calendar schedule is in occupied mode, the fan shall be turned on and after 300 seconds, the heating valve, outside air dampers and cooling coil shall be controlled in sequence to maintain space temperature at 72°F.

| System: Air-handler 1 | Point designation | Device number | Inputs - Analog | | | | Inputs - Digital | | | Outputs - Analog | | | Outputs - Digital | | | Alarms | | | | Comments |
|---|---|---|---|---|---|---|---|---|---|---|---|---|---|---|---|---|---|---|---|---|
| | | | Temperature | Humidity | Flow | Electric current | Freeze thermostat | Differential pressure | Flow switch | Transducer | Current 4–20 ma | Voltage 0–10 Vdc | Contact | Solenoid valve | Relay | Contact open | Contact closed | Value greater than | Value less than | |
| Outside air temperature | AI AH1 OA T | 1 | X | | | | | | | | | | | | | | | | | |
| Mixed air dampers | AO AH1 MAD | 7 | | | | | | | | X | | | | | | | | | | |
| Filter pressure | DI AH1 FT P | 7 | | | | | | X | | | | | | | | | X | | | Filter change alarm |
| Heating coil | AO AH1 HC | 7 | | | | | | | | X | | | | | | | | | | |
| Freeze thermostat | DI AH 1 FR T | 7 | | | | | X | | | | | | | | | | X | | | Freeze alarm |
| Cooling coil | AO AH1 CC | 7 | | | | | | | | X | | | | | | | | | | |
| Humidifier | AO AH1 HM | 7 | | | | | | | | X | | | | | | | | | | |
| Supply fan on/off | DO AH1 SF | | | | | | | | | | | | X | | | | | | | |
| Supply fan electric current | AI AH1 SF EC | 6 | | | | X | | | | | | | | | | | | 105% | 80% | Fan current high alarm or low alarm |
| Supply air temperature | AI AH1 SA T | 2 | X | | | | | | | | | | | | | | | | | |
| Supply air humidity | AI AH1 SA H | 4 | | X | | | | | | | | | | | | | | 85% | | Supply air high humidity alarm |
| Space 001 temperature | AI AH1 001 T | 3 | X | | | | | | | | | | | | | | | 85 | 53 | Space temp high or space temp low alarm |
| Space 001 humidity | AI AH1 001 H | 5 | | X | | | | | | | | | | | | | | 60% | | Space humidity high alarm |

**Figure 11.9** Control Points and Characteristics

164     Fundamentals of HVAC

The control sequence shall be: heating valve fully open at 0% and going to fully closed at 33%, at 34% the dampers will be at their minimum position of 20% and will move to fully open at controller 66%, the cooling valve will be fully closed until 66% and will be fully open at 100%

**Economizer control:** When the outside temperature is above 66°F, the outside air dampers shall be set back to minimum position of 20%, overriding the room controller requirement.

**Fan Control Alarm:** If the fan has been commanded on for 30 seconds, and the fan current is below alarm setpoint 85% of commissioned current, the fan shall be instructed to stop, outside air dampers closed, and heating and cooling valves closed. An alarm of 'low fan current' shall be issued.

If the fan has been commanded off for 10 seconds, and the fan current is above the low limit, the fan shall be commanded off, and dampers, heating coil and cooling coil shall be controlled as in occupied mode. An alarm of 'fan failing to stop' shall be issued.

**Filter alarm:** If the filter pressure drop exceeds 0.3 inches water gauge, the filter alarm shall be issued.

**Freeze Alarm:** If the supply air temperature drops below 45°F, hardware freezestat operates, system changes to unoccupied mode and issues 'freeze' alarm.

**Manual override:** If the manual override is sensed, run in 'occupied mode' for 3 hours.

**System status:** 280 seconds after entering 'occupied mode' the room temperature, supply temperature, and ambient temperature shall be recorded along with current date and time.

Note that, in this case, the point names are given in full. It really helps future maintenance if a point naming convention is established and enforced, including having the contractor label every input and every output device with its point name. It also discourages the contractor from accidentally dropping into the naming convention of the last project!

*The convention used here is only an example,* chosen to make this text easy to understand. Many naming conventions do not include the spaces and many do not include the AI, AO, DI, DO but make the name self explanatory. For example, instead of AO AH1 CC, they might use AH1 CCV, meaning AH1 Cooling Coil Valve.

The column 'Device number' refers the contractor to the specification for the device. In this case, device number 1 is an outdoor air temperature sensor, device number 2 is a duct temperature sensor, and device number 3 is a room temperature sensor.

Most of the sequence of operations shown here can be achieved with any control system. Two DDC specific routines have been included, to aid maintenance and to help avoid energy waste:

The first is to start the fan leaving all controls alone. The fan will circulate air from the space, so after 280 seconds the sensors should have stabilized The space temperature sensor should record the same temperature as the supply air temperature, except for the small rise in temperature due to fan energy that occurs as the air goes through the fan. Lets suppose this rise is normally 1°F on this example system.

One cool day, when the chilled water system is shut down, the operator checks the startup temperature rise. It is surprising to note that it is minus 4°F, so something has gone wrong. It is cool outside, so the outside dampers could be letting in cold air, even when they are controlled to be fully closed. It is also possible that the space temperature sensor or supply air temperature sensor is providing the wrong reading. The operator does not know which is the actual problem but will probably start by checking the dampers.

The erroneous temperature difference will provide different possibilities for what is wrong under different weather conditions, depending on whether the chilled water was available, and whether the temperature difference was positive or negative. The designer can fairly easily work up a written decision tree of possible problems to help the operator. As all the information is available in the DDC system, the designer can also program the system to work through the decision tree and present the operator with the possible problems to check.

This level of sophistication is becoming available on factory produced standard products. On larger systems, and for remotely monitored sites, this type of self-analysis is becoming a valuable feature of high-level DDC systems. However, it is generally not warranted on a small, simple system where the programming is being written for that one project.

The second specific DDC feature is using a current sensor on one of the cables to the fan to provide a measure of fan current. Our example is a constant volume system, so the load on the fan will be relatively constant. It will not be completely constant, since the pressure drop across the filters will rise as the filters become loaded with dirt. Based on the actual fan current when the system is commissioned, a high alarm point and a low alarm point can be chosen. Then, if a bearing starts to fail, the load will typically rise, and this can be detected before bearing failure and possible destruction of the fan. Also, if the fan is belt driven and the belts slip or break, the fan current will drop substantially. This will also be detected. Finally, if the fan starter or motor fails, there will be no motor current, again sensed as low current and alarmed.

These are just two examples of how a small change in how the DDC controls are arranged can provide for better control and maintenance.

Now that we have considered the basics of DDC and a sample system we will move on to how systems are interconnected and built up into networks serving a whole building or many buildings.

## 11.6 Architecture and Advantages of Direct Digital Controls

So far we have considered the controls of a single, simple system connected to a single DDC panel. In many buildings, there will be several systems, often with many more points controlling air-handlers, VAV boxes, heating valves, pumps, boilers and chillers. Wiring from a single huge DDC panel is not a practical option for two reasons. First, failure of the unit means failure of the entire system, and secondly, the wiring becomes very extensive and expensive. Instead, the system is broken down into smaller panels that are linked together on a communications cable, called a "communications network."

It sounds simple, and it is if the system uses equipment from only one manufacturer. However, when more than one manufacturer is involved, it is not as simple. There are three communication issues that create problems. Let us identify them in terms of human communication first.

### Languages

The problem is very similar to the problems of human language. In order for people from different countries to communicate, interpretation or language translation is required.

Similarly, in the controls world, different companies have worked up different languages. The languages differ both in terms of the words and in terms of sentence structure. There are two ways of enabling communication so that one manufacturer's equipment can communicate with another manufacturer's equipment that uses a different programming language. The first is to have an interpreter, called a "**gateway**," between the two units. The second way is to program an additional, common language into both manufacturers' units.

### Vocabulary and Idea Complexity

Different people learn different sets of words in the same language. For simple, everyday things, like bread and water, everyone learns the words in each language. In addition, different people are trained in different skills. Consider, for example, when an engineer and an accountant want to discuss the long-term value of a project. They can find themselves having great difficulty communicating, because they have different vocabularies and different thinking skills in the same language.

### Transmission Method and Speed

Finally, people send messages over long distances by a variety of methods at various speeds. For example, consider a letter being faxed to a remote recipient. It first goes through the fax machine (gateway) to be converted into telephone data. The telephone data is routed through various telephone exchanges (routers) till it reaches the receiving fax machine (gateway) that converts the data back into the original text letter.

In addition to the method, there is an issue of speed. Faxing is a quick and easy way of sending a letter, but if a whole book of text is to be sent, the much higher speed available on the Internet is considerably more attractive.

The issues of language, vocabulary and idea complexity, and transmission method and speed are very much the same in DDC systems.

Typically, a DDC panel includes software that provides the sequence of control activities and software for communicating with other panels. The internal software is generally proprietary to each manufacturer, and the communications software can be proprietary or public. There are several good, reliable communication languages, called "protocols" for simple information such as 'the temperature is 100F', 'open to 60%'. The problems arise as soon as higher level communications, including any form of logic, are required.

In an attempt to eliminate the cost and challenges of no communication or expensive and limited gateways, ASHRAE produced a communications standard called "**BACnet.**" This is a public communications protocol that is designed to allow communication at all levels in a DDC system. It is documented in ASHRAE Standard 135-2004 A Data Communication Protocol for Building Automation and Control Networks[2].

BACnet is particularly aimed at facilitating communications between different vendors' products at all levels. This allows buyers to have more vendor choice. It is important to note, though, that while the BacNet standard establishes rules, the designer still has to be very careful, since the number of rules used by different manufactures can make 'BACnet compatible' systems and components unable to communicate. However, with careful specification, one can obtain units and components from a variety of manufacturers that will communicate with each other.

The ability of different manufacturers' equipment to work together on a network is called "**interoperability**." To assist in ensuring interoperability and the use of BACnet, a BACnet interoperability association has been formed to test and certify products.

### System Architecture

Let us now consider a DDC system and how it might be arranged—the system architecture. Consider the system illustrated in *Figure 11.10*.

Across the top of the figure is a high-speed network connecting main standalone panels and the operator terminal. In this example, the standalone panel on the left uses a different communication protocol (language) from the protocols used by the other two panels and the operator workstation. Therefore, a gateway (translator) connects the standalone panel on the left to the network. A "gateway" is a processor specifically designed to accept specific information in one protocol and send out the same information in another protocol.

Note that gateways are specific in terms of 'protocol in' and 'protocol out' and are often not comprehensive. By "not comprehensive," we mean that only

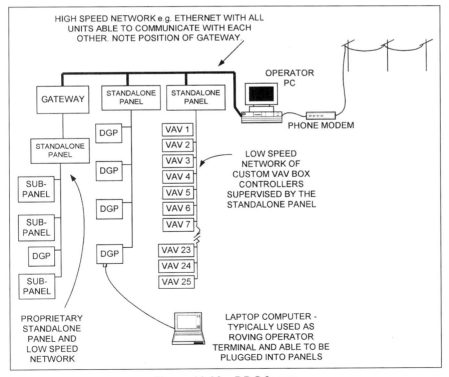

**Figure 11.10** DDC System

specifically chosen information, not all information, can be translated (think of it as a translator with a limited vocabulary and limited intelligence).

The standalone panel on the left has a lower speed network of devices connected to it. The sub-panels might be small processors dealing with an air-handling unit, while the "**data gathering panel**" **DGP**, may be simply gathering outside temperature and some room temperatures and transmitting them to the other panels.

The central standalone panel does all the processing for its branch of the system, with remote DGPs to collect inputs and drive outputs. A laptop is temporarily connected to one of the DGPs to allow the operator/maintenance staff to interrogate the system. The use of a laptop allows the operator/maintenance staff to have access to every function on that network branch, but it may not allow access through the standalone panel to the rest of the system.

The right-hand standalone panel is shown as having numerous VAV box custom controllers connected to it. These controllers are factory-produced, with fixed software routines built in to them. Programming involves setting setpoints and choosing which functions are to be active. These custom controllers are attractive because they are economical, but they are restrictive, in that only the pre-written instructions can be used.

In *Figure 11.10* there are a variety of devices in various arrangements with an operator PC as the local human interface. In addition, a phone "modem" is shown allowing communications with the system via a telephone from anywhere in the world. The modem is a device that converts the digital signals from the PC to audio signals, to allow them to travel on the telephone lines. There are three strikes against modems: they are slow, telephone charges can be prohibitive, and only one connection can be made to the modem.

These restrictions are now being removed by adding a "**web server**." A web server is another computer! The web server connects between the high-speed network and the Internet. It is programmed to take information from the DDC system and to present it, on demand, as web pages on the Internet. This enables anyone who has the appropriate access password to access the system, via the Internet, from anywhere in the world, at no additional cost.

Within the facility, web access allows any PC with web access to be used as an operator station, instead of only specifically designated operator stations. This is much more flexible than having to go to the operator's terminal to access the system. For example, the energy manager can use an office PC to access energy data on the machine that is used for normal day-to-day office work.

This chapter has done no more than introduce you to some of the basics and general ideas of DDC. The system has *advantages* including:

- Increased accuracy and control performance
- System flexibility and sophistication that is limited only by your ingenuity and the available financial resources.
- The system ability to store knowledge about the internal behavior over time and to present this information in ways that assist in energy saving, monitoring, and improved maintenance.
- Remote access to the entire system to modify software, alter control settings, adjust setpoints and schedules via phone or via the Internet.
- With increased use and the falling price of computer systems in general, DDC is often less expensive than conventional controls.

Then, there are the *disadvantages*:

- DDC systems are not simple. Qualified maintenance and operations people are critical to ongoing success. They must be trained so that they understand how the system is designed to operate.
- Extending an existing system can be a really frustrating challenge due to the frequent lack of interoperability between different manufacturers' products and even between upgrades of the same manufacturer's products.

For fairly detailed information on the specification of DDC systems ASHRAE Guideline 13-2000 *Specifying Direct Digital Control Systems4* is available.

## The Next Step

In Chapter 12 we move on to consider energy conservation. We will review the subject in general before a brief discussion of the ASHRAE/IESNA *Standard 90.1-2001 Energy Standard for Buildings Except Low-Rise Residential Buildings* and some heat recovery and evaporative cooling energy saving methods.

## Summary

Chapter 11 has been an introduction to the ideas behind controls. This is a vast field and we have only provided a glimpse of the subject. A more technical and detailed introduction to controls is available as a Self-Study Course in this series *Fundamentals of Controls*.

The chapter started off with some general discussion on control types: self-powered, electric controls, pneumatic controls, electronic (analogue electronic), and direct digital controls. Each of these types has a niche where it is a very good choice but there is a general trend towards DDC controls. We then considered a very simple electric control of a two-element hot water heater to show how controls can be considered in a logical way. Next we introduced the control loop and the difference between open loop control (no feedback) and closed loop control (with feedback). The parts of a control loop that you should be able to identify are: setpoint, sensor, controller, controlled device, and controlled variable.

To illustrate the main issues with modulating controls, we had you imagine pouring water into a glass. The ideas illustrated were

**Proportional control** is control in which the control action increases in proportion to the error from the setpoint
**Offset** is the change of apparent setpoint as the control action increases in a proportional controller
**Gain** the ability of the controller to make a large change in control signal
**Overshoot** is the result of applying too large a control signal and being unable to reduce it in time to prevent overshooting the control point
**Speed of reaction** is the time it takes for the controller to initiate a significant change

Having considered the standard control loops we went on to consider the four main types of DDC points:

**Digital/Binary Input**: a circuit such as a switch closing
**Digital/Binary Output**: providing power to switch a relay, motor starter, or two-position control valve.
**Analogue Input**: typically a signal from a temperature, pressure or electric current sensor.
**Analogue Output:** providing a variable signal to a valve, damper or motor speed controller, often via a transducer that changes the low power signal to a pneumatic or electric power source with the necessary power to drive the valve or damper.

Having introduced the four main point types, we introduced the concept of using a point identification scheme, then we considered a very simple example of a sequence of operations which are the logical instructions for the DDC controller to execute, to provide the required control. The required information to specify a DDC system control was then illustrated with a single air handler. The list of control points and schedule of operations is always required, but the schematic can be omitted, though doing so can lead to misunderstandings at the time of installation.

As well as accuracy, a major advantage of DDC is the ability to record data and either use it for more intelligent control or as information for the operator.

Finally we considered DDC architecture, introduced BACnet and interoperability and listed the pros and cons of DDC.

## Bibliography

1. ASHRAE Self Directed Learning Course *Fundamentals of Controls*
2. ASHRAE Standard 135-2001 *A Data Communication Protocol for Building Automation and Control Networks*
3. ASHRAE/IESNA *Standard 90.1-2001 Energy Standard for Buildings Except Low-Rise Residential Buildings*
4. ASHRAE Guideline 13-2000 *Specifying Direct Digital Controls Systems*

Chapter 12

# Energy Conservation Measures

## Contents of Chapter 12

Study Objectives of Chapter 12
12.1 Introduction
12.2 Energy Considerations for Buildings
12.3 ASHRAE/IESNA Standard 90.1
12.4 Heat Recovery
12.5 Air-Side and Water-Side Economizers
12.6 Evaporative Cooling
12.7 Control of Building Pressure
Bibliography

## Study Objectives of Chapter 12

There are three primary objectives in Chapter 12:
   First we will introduce you to some basic ideas about energy conservation.
   The second objective is to introduce you to ASHRAE/IESNA Standard 90.1 2004 Energy Standard for Buildings Except Low-Rise Residential Buildings[1]. (Standard 90.1) This standard, produced cooperatively by ASHRAE and the Illuminating Engineering Society of North America, is becoming the minimum standard for new buildings in the USA.
   Lastly, we are going to look at four ways that HVAC systems can be designed to use less energy.
   After studying the chapter, you should be able to:

Explain energy conservation and some basic ways of thinking about it.
Describe, generally, the contents of Standard 90.1.
Describe the equipment and operation of the heat wheel, heat pipe and runaround methods of heat recovery.
Describe the process and be able to provide examples of uses of evaporative cooling.
Explain the significance of building pressure.

## 12.1 Introduction

During this course we have mentioned and discussed the differences between initial cost and cost-in-use that are relevant to various types of equipment. In many instances, the savings on the initial cost of equipment is squandered because the equipment is more expensive to run, due to excessive energy costs that are incurred over the life of the building.

The objective of energy conservation is to use less energy. This is accomplished by various methods, including recycling energy where useful. Energy conservation should be part of the entire life cycle of a building: it should be a consideration during the initial conception of a building, through its construction, during the operation and maintenance of the building throughout its life, and even in deconstruction.

It is important for everyone who participates in the design, operation and maintenance of the building to realize that, however energy efficient the system as initially designed and installed, the energy efficiency will degrade unless it is operated correctly and deliberately maintained.

In order to improve the energy performance of buildings and provide a benchmark for comparison ASHRAE/IESNA has issued Standard 90.1 Energy Standard for Buildings Except Low-Rise Residential Buildings. The Standard sets out minimum criteria for the building construction and mechanical and electrical equipment in the building and we will discuss it later in the chapter.

## 12.2 Energy Considerations for Buildings

The energy consumption of a building is determined from the very first design decisions through to final demolition.

### Conception and Design

In the very beginning of the design process, many architectural choices can be made to significantly increase, or decrease, the energy consumption of a building. For example, large un-shaded windows that face the afternoon sun can greatly increase the cooling load. Alternatively, the same windows, facing north produce a relatively small cooling load.

It is at the early design stage that the mechanical designer should become seriously involved in the building design as a whole. Historically, the architect would design the building, and then send a set of plans to the mechanical designer to design the HVAC. This model does not work well to produce energy-efficient buildings, because many early design choices can facilitate energy conserving design or make them totally impossible or uneconomic.

### Consider this Example:

In cold climates, a perimeter hot water heating system is often used to offset the heating losses through the wall and windows. Because modern windows are available with insulation values, that approach the insulation value of traditional walls, if the architectural design specifies walls and windows with higher insulation values, the perimeter heating system requirements could be avoided. However, this is a suggestion that would typically be made by the mechanical designer, and the choice can only be made very early in the project. If the mechanical designer suggests a more energy efficient design, this could have a negative impact on the mechanical design fee. Why? Building owners often contract with the design team members for a fee that is based on a percentage of their individual portion of the building cost. In the example just given, the fee for the mechanical designers would include a percentage of the cost of the perimeter heating system. As a result, if the mechanical designers suggest that the perimeter heating be omitted in favor of higher priced windows, the they could be forfeiting a substantial portion of their fee. Hardly an incentive to the engineer to suggest the idea!

Since this method of calculating the mechanical design fee does not encourage energy conservation, what other alternatives are available?

Imagine an alternative fee structure, where the total design fee for the mechanical design would be calculated as a percentage of the cost of the completed building, rather than of the specific mechanical design elements. Then, the mechanical designers could make design suggestions that would not have a negative impact on their design fees. Furthermore, imagine what would happen if the contract also specified that an objective of the building design included energy savings, and provided the entire design team with financial bonuses based on achieving the energy savings. Then the design team would have an incentive to spend time on designing energy efficient buildings!

How could this bonus incentive be structured? Consider what would happen if the bonus represented half the energy savings that were achieved during the first five years after the building was completed, (based on the estimated energy costs for a conventional building design). In this case, the design team would have an incentive to design for maximum energy savings. The result would be that the operating expense for the owner would be reduced by half the energy cost reduction during the first five years, and after the first five years, the owner would receive the benefit of all future energy-related savings. In this scenario, the owner could save money by setting up the contract to encourage desired behavior! Notice that there is not necessarily any additional capital cost to the owner, only the likelihood of operational cost savings: a huge return based only on some contract wording.

In case you are thinking it would never work, you should know that many owners are willing to contract to have energy conservation specialists come back, after construction is completed, and to pay them a significant fee, in addition to retrofit costs, to fix what could have been achieved as part of the original design at a fraction of the cost. We will discuss energy conservation that can be achieved through retrofit in the section entitled: "Turn it in."

## *Construction*

The best possible building plans can be made a mockery by poor construction. If windows and doors are not sealed to the walls, and/or if insulation is installed unevenly and with gaps, the air-leakage can be costly in terms of both energy and building deterioration. The mechanical plant must be installed and set working correctly. Many systems are surprisingly robust, and gross errors in installation can go undetected, making the building less energy efficient —and less comfortable—than it was designed to be.

## *Operation*

If the staff does not know how a system is meant to work, there is a very high probability that they will operate it differently and, more than likely, not as efficiently. It is really important that staff are taught how the systems are designed to work and provided with clear, easy to understand, written instructions for later reference. A pile of manufacturer's leaflets may look pretty but it does not explain how all the bits are meant to work together.

### Maintenance

With limited maintenance, even the best equipment will falter and fail: Controls do not hold their calibration and work indefinitely. Control linkages wear out; damper seals lose their flexibility; cooling towers fill up with dust; the fill degenerates; and chiller tubes get fouled with a coating which reduces their heat transfer performance. The list of maintenance requirements is very long, but critical for maintaining energy-efficient building performance.

### Three Ways to Save Energy

The mantra of energy savings is: Turn it off. Turn it down. Turn it in.

#### Turn it off

This is the simplest and almost always, the least expensive method to implement, and it has the highest saving. If a service is not required, can it be turned off? There are usually several alternatives that can be considered to shorten the running time to the minimum.

Opportunities to "Turn it off" can be found at the design phase and at the operational phase of a building's life cycle.

Let us take a simple example of stairway lighting in a mild climate. For this example, we will ignore any local issues of safety or legislation:

A four-storey apartment building has stairs for access. If the stairs are fully enclosed, the lights must be "on" all the time for people to see their way up and down the stairs.

The first alternative for energy savings can be identified early in the design phase of the building: Designing the stairs with large windows allows the lights to be turned off during daylight hours. The light switching can easily be done with an astronomical clock, or better still, a photocell. The astronomical clock allows for the changing lengths of the day, while the photocell senses the light level and switches on and off at a preset light level.

At both the design and the operation phase of the building's life cycle, a second savings opportunity exists. To discover it, consider asking the question: "What is the objective of having the lights on?" The lights are to provide illumination for people to go up and down the stairs. The next question is, "Is there a way to provide illumination when it is required, and yet not have the lights on when it is not required?" Several solutions come to mind. A low tech solution could be the installation of a pneumatic push-button timer switch at each level. Then, people entering the stairwell could push the button and turn the lights on for, say, ten minutes. The advantage is that, now, we have a simple system that provides the required service when it is required. However, there is an education requirement with a system like this. People need to be shown where the light switch is located. And they need to be taught that, even if the stairwell has been illuminated because an earlier person turned the switch "on," they still have to reactivate the switch, in order to provide continuous illumination while they are in the stairwell. For example, if one person has entered the stairwell and depressed the switch, the stairwell will be illuminate for ten minutes. Nine minutes later, a second person, enters the stairwell and, because the light is "on," does not look for a switch. While that second person is in the stairwell, the lights will go off, leaving that person in the dark. As a result, graphics-based signage would be required, to manage issues based on language and reading skills. Therefore, to alleviate these signage issues, as an alternative, the switch could be wired to detect and respond to the opening of the lobby door or motion detectors could be used to turn the lights "on."

The above example illustrates how a building design choice, in this case, windows, allowed a substantial reduction in operating hours. Then thinking about "What is the objective?" allowed a further, large, reduction in operating hours.

Determining design parameters based on a requirement to "turn it off" may seem extreme, but it is the norm in many parts of the world. You would probably be surprised at how many opportunities you could find in your own experiences where things could be turned off, and energy could be saved, if the focus was on providing only what is needed.

Now let's go on the second approach, which tends to be more complicated, and therefore more costly, to work out and implement.

*Turn it down*
"Turn it down" meaning reduce the amount of heating, cooling or other process while still providing the required service. In Chapter 4, when we covered $CO_2$ control of ventilation air, we discussed the idea of only providing the required amount of a service at the time it is needed. As you recall, $CO_2$ was used as a surrogate (indicator) for assessing the room population and deciding how much outside air was required for the current occupants. Using $CO_2$ as a surrogate allowed the amount of outside air to be turned down when the room population was low.

There are numerous examples of using "turn it down" as an energy conservation tool. Two that are commonly implemented include:

**Heating reset**: In Chapter 8 we discussed resetting the heating water temperature down, as the load drops. This reduces piping heat losses and improves control. However, on a variable speed pumping system, lowering the water temperature increases water volume required and so increases pumping power. The issue is finding the best balance between temperature reset and pumping power.

**Chilled water temperature reset**: The chilled water system is designed for the hottest and most humid afternoons that happen a few times a year. The rest of the time the chilled water system is not running to full capacity. Except in a very humid climate, where dehumidification is always a challenge, the chilled water temperature can probably be reset up a degree or two or more. This improves chiller performance and generally saves energy.

*Turn it in*
"Turn it in" means "replace with a new one." This is the third way of saving energy. It is almost always the most difficult to justify, since it is the most costly. For example, your building may have a forty-year-old boiler with a seasonal efficiency of only 50%. A modern boiler might raise the seasonal efficiency to 70% and provide a fuel saving of 28%. Although the percentage saving is substantial, it can be frustrating to find that it would take 12 years to pay for a new boiler out of the savings. Typically, a 12-year payback is too long for the financial officer to accept.

It almost never pays in energy savings to <u>replace</u> building fabric. For example, replacing single pane windows with double or triple pane or replacing a roof with a much better insulated roof usually have energy savings that pay for the work in 30 years or more. However, if the windows are going to be replaced because they are old and the frames have rotted, then it almost always worth spending a bit extra on a higher energy-efficient unit. Here, one is comparing the extra cost of better windows against the extra energy savings, and it is usually an attractive investment.

While it almost never pays to replace building fabric, we should also note that it is usually economically worthwhile to repair the building fabric, particularly where there are air holes. For example, many industrial buildings have concrete block walls up to the roof. Over time, the block walls may well drop a bit, leaving a gap between wall and roof. Plugging this gap with expanding foam is a simple task and can reduce the uncontrolled flow of air into, and out of, the building. In a humid climate, this can substantially reduce the dehumidification load; in a cool climate, it could provide substantial heating energy saving.

It is exactly the same for the plant. The boiler may be 40 years old but it will work better if the burner is regularly serviced.

Chillers are another area of consideration. Due to the regulated phase-out of CFC refrigerants, many owners are being forced to consider chiller replacement. If the chiller is to be replaced anyways, it is worth taking the time to calculate the extra savings that are available from a high efficiency unit as compared to the extra cost for the unit. It is highly likely that the difference in cost for the high efficiency chiller will have a speedy payback in energy savings.

Having introduced three ways of saving energy – Turn it off – Turn it down – Turn it in, let's move on to a Standard that sets minimum requirements for energy saving in new buildings and major renovations.

## 12.3 ASHRAE/IESNA Standard 90.1

ASHRAE and the Illuminating Society of North America (IESNA) wrote ASHRAE/IESNA Standard 90.1 *Energy Standard for Buildings Except Low-Rise Residential Buildings* (Standard 90.1) as a joint venture. The latest printed edition is 2004, which was used for this text. There is a detailed, well-illustrated, and explanatory companion document, 90.1 *User's Manual ANSI/ASHRAE/IESNA Standard 90.1-2004 Energy Standard for Buildings Except Low-Rise Residential Building*[2].

The purpose of the Standard is "to provide minimum requirements for energy-efficient design of buildings except low-rise residential buildings." It is a <u>minimum</u> standard and there are some energy reduction programs such as "**Leadership in Energy and Environmental Design, LEED,**" that encourage designs to have a lower energy cost than the Standard prescriptive cost. Note that the LEED program gives no acknowledgement unless design energy cost is at least 15% below the Standard 90.1 requirements.

The Standard 90.1 requirements can be met by either complying with all "Prescriptive and Performance Requirements" or by producing a design that has no higher energy cost in a year than a prescribed calculated "Energy Cost Budget."

## Prescriptive and Performance Requirements

The Standard is divided into sections that often fall to different designers. The first section of the Standard is the "Administration and Enforcement" section, to help designers and code officials. It then has six prescriptive sections that define the performance of the components of the building. Finally, it concludes with a calculation method, the "Energy Cost Budget Method" section.

The following is a brief introduction to the sections.

## Building Envelope

The objective of the Standard is to ensure that design choices are both energy-efficient and cost-effective. Therefore, for example, the insulation requirements are more demanding in the colder climates.

The Standard divides climates according to temperature and moisture conditions. The temperature divisions range from the continuously hot, with no heating demands, through to the continuous heating with no cooling requirements. The designer chooses the temperature range relating to the building location, and, on a single page finds the thermal transmission requirements for the building fabric: roofs, walls, floors, doors and fenestration (windows). This is the section for the architect!

The Standard requires slightly higher performance for residential buildings, since they are generally in operation 24 hours of every day. In comparison, many non-residential buildings are in full operation for less than half the hours in a week.

One of the major problem areas of modern buildings is the sealing around penetrations in the building envelope. The building envelope includes the entire perimeter of the building: the windows, doors, walls, and the roof. The allowable leakage around windows and doors is defined. All other parts of the building envelope are covered by the hope-filled request: "The following areas of the building envelope shall be sealed, caulked, gasketed, or weather stripped to minimize leakage." In order to reduce the likelihood of future problems, it is worth the effort to ensure that the contractor fulfills this as a requirement.

The Standard allows some trade-off between the various sections of the building envelope as long as the required overall envelope performance is maintained.

## Heating, Ventilating, and Air conditioning

For single zone buildings of less than 25,000 ft$^2$ and only one or two floors, there is a simplified approach, due to the limited number of choices that designers can make for equipment. As long as the building is a single zone, with one unit, the code requires that the unit will comply with a few straightforward energy saving requirements.

For larger buildings there are numerous requirements for minimum equipment efficiencies in terms of Energy Efficiency Ratio, "**EER**," Coefficient of Performance, "**COP**," and Integrated Part-Load Value "**IPLV**." The following section explains the meaning of each of these terms.

**EER Energy Efficiency Ratio** is the ratio of net cooling capacity in **Btu/hour** to electrical input in **Watts**. A small window air-conditioner, for example, is required to have a minimum EER of 9.7. This is the same as saying that it will provide 9.7 Btu/hour of cooling for an input of 1 watt, under specific test conditions. A watt is 3.412 Btu/hour so the EER of 9.7 requires 9.7 Btu/hour cooling for 3.412 Btu/hour energy input. This works out to about 2.84 times as much cooling energy as compressor energy.

The requirements for water chillers are given in IPLV and COP.

**IPLV, Integrated Part-Load Value** is a weighted average value of EER based on full and part load performance and is used instead of EER on larger electrically driven air-conditioners.

**COP, Coefficient Of Performance**, is the heat removal to energy input in consistent units. For air-cooled chillers, the minimum requirement is COP of 2.8. However, a water cooled centrifugal chiller over 300 tons, has a required minimum COP of 6.0, twice the cooling capacity per watt of the air-cooled machine. This is an area where judicious choice of equipment can make large differences in energy consumption.

In Chapter 10.1, we discussed the statement that "big plant is more efficient." In the case of chillers, this is very true. Unfortunately, COP efficiency is not the only relevant consideration. Other energy inputs for the central plant include the energy for pumping the chilled water to end use and the condenser water to the cooling tower. In addition, the distribution-pipe heat gains must be deducted from the cooling capacity.

Having defined minimum equipment performance, the Standard then goes on to establish rules about controls and installation including insulation, system balancing and commissioning, to ensure minimum equipment utilization efficiency. We have already discussed some of the controls requirements in the previous chapter.

### Service Water Heating

The section on service water heating covers minimum equipment performance and maximum standby loss. Also detailed are pipe insulation and recirculation requirements.

### Lighting

On average, in the USA, buildings use about 35% of their total energy for lighting. This provides a big opportunity for savings. The Standard allows a specific number of Watts per square foot, $W/ft^2$, however, the designer is given a certain amount of leeway in the calculations: The allowed $W/ft^2$ can be calculated on the basis of type of building or on a room-by-room basis. The Standard allows trading between areas and between lighting and HVAC, as long as the net energy cost through the year is not increased above the prescribed allowance.

The Standard recognizes variation in use of the same type of space in different types of buildings. So, for example, corridors generally have an allowance of 0.5 $W/ft^2$ but this is raised to 1.0 $W/ft^2$ for hospitals.

### Energy-Cost Budget Method

The energy-cost budget is a way to allow designers to have the flexibility to design the building according to their needs, as long as it does not cost more in energy than the Standard permits. To use the Energy-Cost Budget Method, the designer is instructed to calculate the energy-cost budget for standard plant equipment, then to compare that to the cost of the energy required by the equipment chosen.

The Energy-Cost Budget, ECB, requires the use of hour-by-hour building energy analysis software. No particular software is specified, but software performance is mandated. Local utility rates are used in the simulation. The building has to be analyzed, using the prescribed building envelope and equipment efficiencies, to obtain the 'energy-cost budget' and again with actual envelope and equipment. Compliance is achieved when the 'design energy-cost' does not exceed the 'energy-cost budget'.

If you become involved in using Standard 90.1, remember that the User Manual provides a clear, easy-to-follow explanation of how to use and apply the Standard.

## 12.4 Heat Recovery

When designing to comply with the Standard, designers can minimize energy use by reducing the energy requirements of a building, and/or by energy recovery. During design, always aim first to minimize energy use before considering energy recovery. The reason is that heat recovery is almost always involved with "**low-grade heat**." Low-grade heat is heat that is at a temperature relatively close to the temperature at which it can be used at all. Low-grade heat requires oversized heat transfer surfaces and can often only fill a part of the load. For example, the condenser water from a chiller at 95°F can be used to preheat domestic service water to 90°F but no hotter. A valuable contribution, but it does not do the whole task, since 140°F is the typical requirement.

There are cases where systems can be deliberately chosen to integrate with low heat sources. A good example of this is the use of **condensing boilers** with radiant floor heating systems. The **condensing boiler** is a boiler with an additional flue gas heat recovery section. In this additional flue gas cooling section, the water vapor in the flue gas is condensed, causing it to give up its latent heat. This increases the boiler efficiency from a maximum of about 85%, with a flow temperature of 180°F, to about 95% with a 105°F flow temperature. Since radiant flooring operates at low water temperature, the condensing boiler is an excellent match for the radiant floor. Note that condensing only begins to occur below 135°F, so buying a condensing boiler and running it near 135°F will reduce the boiler efficiency since it will not condense the flue gas water vapor.

### *Energy Recovery Coils: Run-Around Coils*

One way to achieve energy recovery is with run-around energy recovery coils. A typical run-around coil arrangement is shown in *Figure 12.1*.

In summer, the conditioned exhaust air cools the fluid in the exhaust air coil. This fluid is then pumped over to the supply air coil to pre-cool the incoming outside air.

In winter the heat transfer works the other way: the warm exhaust air heats the fluid in the exhaust air coil, which is then pumped over to the supply air coil to heat the cold incoming air.

At intermediate temperatures the system is shut off, since it is not useful.

When outside temperatures are below freezing, the three-way valve is used with a glycol anti-freeze mixture in the coils. In cold weather, some of the fluid

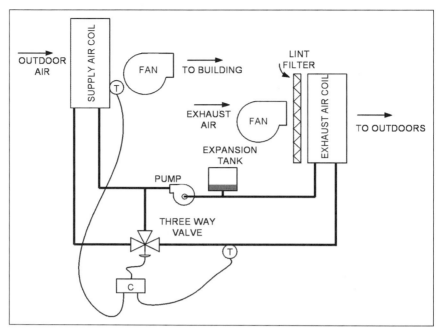

**Figure 12.1**  Run-Around Energy Recovery Coils

bypasses the supply air coil, to avoid overcooling. The mixture of very cold fluid from the supply air coil and diverted fluid, mix to a temperature that is high enough to avoid causing frost on the exhaust air coil. The maximum amount of cooling that can be achieved with the exhaust air coil is limited by the temperature at which frost starts to form in the coil. This frosting of the exhaust coil effectively sets a limit to the transfer possible at low temperatures.

In *Figure 12.1*, a filter is shown in front of the exhaust air coil. It is important to include this filter, since omitting it will soon cause a clogged coil. This is particularly true if the coil runs wet with condensation in cold weather.

The run-around coil system has three particular advantages.

1. There is no possibility of cross contamination between the two air streams. This factor makes it suitable for hospital or fume hood exhaust heat recovery.
2. The two coils do not have to be adjacent to one another. A laboratory building could have the outside air intake low in the building and the fume hood exhaust on the roof, with the run-around pipes connecting the two coils.
3. Run-around coils only transfer heat, not moisture.

### Heat Pipes

A heat pipe is a length of pipe with an interior wick that contains a charge of refrigerant, as shown in *Figure 12.2*.

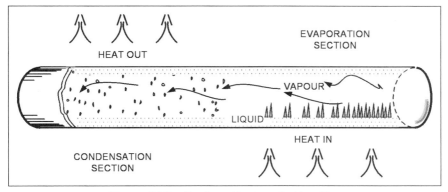

**Figure 12.2** Cut Away Section of a Heat Pipe

The type and quantity of refrigerant that is installed is chosen for the particular temperature requirements. In operation, the pipe is approximately horizontal and one end is warmed, which evaporates refrigerant. The refrigerant vapor fills the tube. If the other half of the tube is cooled, the refrigerant will condense and flow along the wick to the heated end, to be evaporated once more. This heat-driven refrigeration cycle is surprisingly efficient.

The normal heat pipe unit consists of a bundle of pipes with external fins and a central divider plate. *Figure 12.3* shows a view down onto a unit that is mounted in the relief and intake air streams to an air-handling unit. Flexible connections are shown which facilitate the tipping. To adjust the heat transfer, one end or the other end of the tubes would be lifted.

The outside air is cold as it comes in over the warm coil. This warms the air, and the tube is cooled. The cooled refrigerant inside condenses, giving up its latent heat, which heats the air. The re-condensed refrigerant wicks across to the exhaust side and then absorbs heat from the exhaust air. This

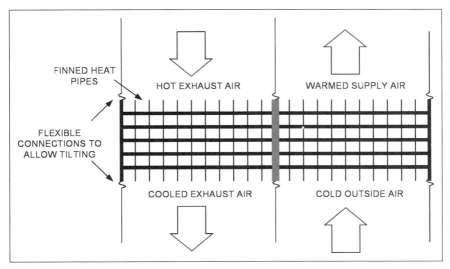

**Figure 12.3** Heat Pipe Assembly in Exhaust and Outside Air Entry

heat evaporates the refrigerant back into a vapor which fills the pipe, and is again available to warm the cold outside air.

The usual heat-pipe unit must be approximately horizontal to work well. A standard way to reduce the heat transfer is to tilt the evaporator (cold) end up a few degrees. This tilt control first reduces, and then halts, the flow of refrigerant to the evaporator end, and the process stops.

*Figure 12.3* was based on winter operation. In summer, the unit only has to be tilted to work the other way and cool the incoming outside air as it heats the outgoing exhaust air.

The unit is designed as a sensible heat transfer device, though allowing condensation to occur on the cold end can transfer worthwhile latent heat. Effectiveness ratings range up to 80% with 14 rows of tubes. However, each additional row contributes proportionally less to the overall performance. As a result, the economic choice is ten or fewer rows.

A major advantage of the units is very low, to no, cross contamination.

### *Desiccant Wheels*

Desiccants are chemicals that are quick to pick up heat and moisture, and quick to give them up again if exposed to a cooler, drier atmosphere. A matrix, as indicated on the left of *Figure 12.4*, may be coated with such a chemical and made up into a wheel several inches thick. In use, the supply air is ducted through one half of the wheel and the exhaust air through the other half.

Let us suppose it is a hot summer day, so the exhaust is cooler and drier than the supply of outside air. The chemical coating in the section of the coil in the exhaust stream becomes relatively cool and dry. Now the wheel is slowly rotated and the cool, dry section moves into the incoming hot, humid air, drying and cooling the air. Similarly, a section is moving from hot and humid into cool and dry, where it gives up moisture and becomes cooler.

The wheel speed, a few revolutions per minute, is adjusted to maximize the transfer of heat and moisture. Control of wheel speed to truly maximize savings is a complex issue, since the transfer of sensible and of latent heat do not vary in direct relation to each other.

The depth of the wheel is filled with exhaust air as it passes into the supply air stream, so there is some cross-contamination. There are ways of minimizing this cross contamination, but it cannot be eliminated. In most comfort situations, the cross contamination in a well-made unit is quite acceptable.

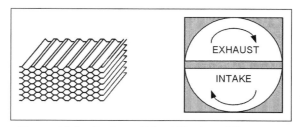

**Figure 12.4** Desiccant Wheel Matrix and Operation

## 12.5 Air-Side and Water-Side Economizers

### *Air-Side Economizers*

In the previous chapters, you have been introduced to the air-side economizer on air-handling units. It is the mixing arrangement that allows up to 100% outside air to be drawn in and relieved in order to take advantage of cool outside air, providing "free cooling." Nothing is free! The air-side economizer equipment costs extra to purchase, there are more components to maintain, and, depending on the climate, the hours when the economizer is actually saving cooling energy may be very limited. In climates that are warm and humid, the number of hours when the outside air has a lower enthalpy than the return air enthalpy may be very few. Thus, Standard 90.1 does not require air-side economizers in most of Florida.

One critical issue with economizers is that their controls must be integrated with the mechanical cooling. This prevents the economizer from increasing the mechanical refrigeration load.

Standard 90.1 has very specific requirements on the control of economizers and, in particular, prohibits the use of mixed air control for economizers on systems that serve more than one zone. Instead, the Standard requires that a supply air thermostat be used to control the cooling coil and economizer. This control method works well as long as the chilled water valve and, if there is one, the heating valve, close fully. If the valves do not close, due to being worn or incorrectly set up, it is possible for the system to use much more energy than expected. Therefore, when this control method is used, it is important that the system be maintained, or that a control sequence is included that will show up the fact that one of the valves is not closing correctly. This control sequence was discussed in section 11.5.

#### Advantages of the air-side economizer

- A low air pressure drop.
- Substantial mechanical-cooling energy savings.
- Reduced water usage in cooling tower systems.

#### Disadvantages of the air-side economizer

- Extra capital cost for the 100% intake and relief air equipment, which includes a return fan on larger systems.
- A higher ongoing electrical operating expense.
- A potential requirement for additional humidification during winter operation.

### *Water-Side Economizers*

The water-side economizer consists of a water-cooled coil, located in the air stream just before the mechanical-cooling coil. The coil can be supplied with water directly from the cooling tower or via a plate heat exchanger. If the water is supplied directly from the tower, the water treatment and cleaning process must be of a high standard, to ensure that the valves and coil do not clog up with dirt. If a heat exchanger is used, there is the additional cost of the exchanger, and the heat transfer will be less efficient, since there has to be a temperature rise across the exchanger for it to work.

184   Fundamentals of HVAC

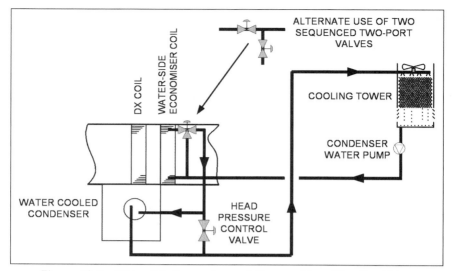

**Figure 12.5**   Water-Side Economizer and Alternate Use of Two-Port Valves

There are several possible arrangements, depending on the particular equipment and sizes. One example for packaged units is shown in *Figure 12.5*. The three-port valve determines how much of the tower water flows through the economizer coil, and the two-port valve determines how much water bypasses the condenser to avoid the condenser being overcooled.

Note that the three-port valve can be replaced with two two-port valves, as shown in the detail. The valves would be sequenced so that, as one opens, the other closes, to provide the same effect as the mixing valve, but often at lower cost in small sizes.

The **"head pressure"** is the pressure in the refrigeration condenser. If the head pressure falls below the required pressure, the valve is opened to reduce water flow through the condenser. On cool days, when the tower produces very cold water, the valve will stay open, since adequate cooling is provided at well below full design flow.

### Advantages of water-side economizers

- Water-side economizers reduce compressor energy requirements by pre-cooling the air.
- Unlike air-side economizers, which need full sized intake and relief ducts for 100% outside air entry or for 100% exhaust, water-side economizers simply require space for two pipes.
- Unlike the air-side economizer, the water-side economizer does not lower the humidity in winter, saving on possible humidification costs.

### Disadvantages of a water-side economizer

- Higher resistance to airflow, therefore higher fan energy costs.
- Increased tower operation with consequent reduction in life.
- Increased water and chemicals cost.

## 12.6 Evaporative Cooling

You have been introduced to the idea of evaporative cooling several times so far in this course. In Chapter 2 the process of using direct evaporation was introduced in connection with the psychrometric chart.

### Direct Evaporative Cooling

The direct evaporative cooler simply evaporates moisture into the air, reducing the temperature at approximately constant enthalpy. In a hot dry climate this process may often be enough to provide comfortable conditions for people.

In medium to wet climates, the increase in moisture content is frequently not acceptable for sedentary human comfort but is considered acceptable for high effort work places and is ideal for some operations, such as greenhouses.

### Indirect Evaporative Cooling

An indirect evaporative cooler uses evaporation to cool a surface, such as a coil, that is then used to cool the incoming air. The indirect evaporative cooler, which reduces both temperature and enthalpy, can be very effective in all but the most extreme conditions. The two processes are shown on the psychrometric chart, *Figure 12.6*.

A previous Figure, *Figure 12.5* showed the indirect cooler as the "water-side economizer," located before the mechanical cooling coil. That is just one arrangement of two-stage cooling.

*Figure 12.7* shows an alternative to this arrangement.

In this indirect evaporative-intake cooler, water flows down the outside of the air intake passages. As it flows down, outside air is drawn up over the water causing evaporation and cooling. The cooled water cools the intake air passages and hence the intake air. This is shown diagrammatically on the left side of *Figure 12.7*. The unit is mounted at the intake to the air-handler as shown on the right hand side of *Figure 12.7*.

Depending on the local climate, a unit like this can reduce the peak mechanical refrigeration by 30% to 70% with a very low water and energy requirement from the indirect cooler. The performance may be improved even further if the relief air from the building is used as the air that passes over the evaporative surface.

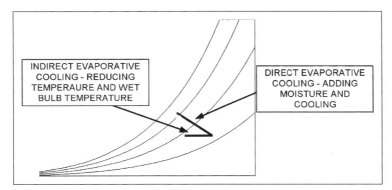

**Figure 12.6** Psychrometric Chart Showing Direct and Indirect Evaporative Cooling

186 Fundamentals of HVAC

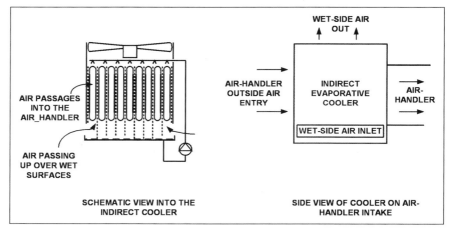

**Figure 12.7** Indirect Evaporative Intake Cooler

To quote from ASHRAE 2000 HVAC Systems and Equipment[3], Chapter 19:

> "Direct evaporative coolers for residences in desert regions typically require 70% less energy than direct expansion air conditioners. For instance, in El Paso, Texas, the typical evaporative cooler consumes 609 kWh per cooling season as compared to 3901 kWh per season for a typical vapor compression air conditioner with a SEER 10. This equates to an average demand of 0.51 kW based on 1200 operating hours, as compared to an average demand of 3.25 kW for a vapor compression air conditioner."

The main advantages of evaporative cooling include:

Substantial energy and cost savings
Reduced peak power demand and reduced size of mechanical refrigeration equipment
Easily integrated into built-up systems

The big disadvantage for evaporative cooling is that many designers don't understand the opportunity!

## 12.7 Control of Building Pressure

Control of building pressure can have a significant effect on energy use, drafts through exterior doors, and comfort. In a hot and humid climate, it is valuable to keep the building at a slightly positive pressure. This ensures that dry air, from inside the building, enters the walls rather than allowing humid air from outside to enter the building through the wall and likely cause mould growth. In a cold climate, the building should be kept close to outside pressure, or slightly negative, to prevent the warm, moist air from inside the building from entering the wall where it could and cause condensation or ice.

Energy Conservation Measures    187

When an economizer is running with 100% outside air, the same amount of air must also leave the building. On small systems, no return or exhaust fan is provided, on the assumption that the washroom exhaust plus leakage will be adequate to balance the amount of air coming in.

In milder climates, intermediate size plants can be accommodated with "**barometric dampers**." Barometric dampers blow open when there is a slightly greater pressure, than outside at that location in the building 'At that location' is included as a proviso, since the wind can make a huge difference to the pressure at different points around a building. If the wind is blowing towards the damper, it will tend to keep it shut. On the other hand if the damper is on the leeward side of the building, the wind will tend to open it.

On the larger economizer systems, typically the ones shown in the figures in this text, complete with a return fan, the return/relief fan and relief damper can be used to control building pressure. The least efficient method is to separately control the relief damper and effectively throttle the relief fan flow. Better, is to add a speed control for the return fan so that it maintains a set minimum outlet pressure. This will ensure adequate return air for the main supply fan and allow the relief damper to control the building pressure.

## The Final Step

Chapter 13 is the final chapter. In it we cover two groups of topics that did not fit into the flow of the previous chapters. The first group deals with heating and heat storage. The second group deals with air distribution in rooms and separate outdoor air systems.

Finally, there are some suggestions for you for future courses and other sources of information.

## Summary

### 12.2 Energy Considerations in Buildings

The objective of energy conservation is to use less energy and to recycle energy where useful. In the design of new facilities it is very important that the whole design team, including the client, have energy conservation as an objective. There is considerable synergy to be gained from a group effort. The client has the ability to set up a design contract that encourages energy conservation to the mutual financial benefit of the team and the client.

There are three ways of achieving energy conservation: **Turn It Off**, turning equipment off, **Turn It Down**, reducing equipment output and **Turn It In**, by replacing equipment with something more efficient. Of these three ways, 'turning equipment off' is usually the most cost effective, with 'turning down' second. Replacement is often not economic.

### 12.3 ASHRAE/IESNA Standard 90.1-2001

To assist in energy conservation ASHRAE/IESNA Standard 90.1-2001 Energy Standard for Buildings Except Low-Rise Residential Buildings was produced, and it is now being adopted in parts of the USA. This standard sets minimum requirements for the building envelope, electrical systems including lighting, and the HVAC, under a prescriptive approach. The HVAC section covers the efficiency of individual equipment, as well as how they are to be interconnected and controlled. In addition, the design team may choose to meet the Standard using the performance route, the Energy Cost Budget Method, in which the design team demonstrate that their design will have no higher energy cost than the prescriptive design would have cost.

The requirements are designed to be easily cost effective and many programs, such as the LEED program, require substantially lower energy consumption than the Standard requires, to be recognized as energy conserving designs.

## 12.4 Heat Recovery

Heat recovery is the reuse of surplus heat from a building, often the exhaust air. Methods of recovering heat from the exhaust were described. These included:

Run-around coils, which are a system where a fluid, water or glycol mixture, is pumped through coils in the exhaust and outside air intake. This transfers heat from the intake air in summer and adds heat to the incoming air in winter. The system has advantages of no cross contamination and the intake and exhaust can be remote from each other just interconnected by the pair of run-around coil pipes.

The heat pipe and desiccant wheels were also described. They both require the intake and exhaust air to pass by each other and have a cross contamination challenge. On the other hand they are often less costly and more effective than the run-around coil.

## 12.5 Air-Side and Water-Side Economizers

The airside economizer is the use of outside air to provide cooling when the outside ambient temperature and humidity can provide 'free cooling'. The system is not economic in very hot humid climates and it creates a low humidity indoors in cold weather.

The waterside economizer uses water, cooled in a cooling tower, to lower the incoming air temperature by means of a pre-cooling coil. The system takes up little space and does not require the large intake duct that the air–side economizer requires. It also has the advantage of not lowering the indoor humidity in cold weather.

## 12.6 Evaporative Cooling

Evaporative cooling can be direct or indirect. Direct evaporative cooling reduces the temperature and raises the humidity by direct evaporation of water in the air. For human comfort, this is a very acceptable situation in a hot dry climate but not useful in a hot and humid climate. For some industrial processes and greenhouses in particular, it can be very effective in all but the most humid climates.

Indirect evaporative cooling uses water that has been cooled by a cooling tower, or by direct evaporation on the outside of a coil, in the incoming air stream. Indirect evaporative cooling lowers both the temperature and the enthalpy. In many climates this can significantly reduce the required size of the mechanical cooling and drastically cut the electrical consumption by lowering the load on the mechanical cooling system.

## 12.7 Control of Building Pressure

If the building pressure is much higher than outside pressure, there will be leakage outwards. Similarly a low inside pressure draws air in through all the building cracks and leaks. Neither over nor under pressure is desirable, as they cause discomfort, energy waste and deterioration of the building fabric.

## Bibliography

1. ASHRAE/IESNA Standard 90.1 2004 Energy Standard for Buildings Except Low-Rise Residential Buildings
2. ASHRAE User's Manual ANSI/ASHRAE/IESNA Standard 90.1 2004
3. ASHRAE 2000 HVAC Systems and Equipment

Chapter 13

# Special Applications

## Contents of Chapter 13

Study Objectives of Chapter 13
13.1 Introduction
13.2 Radiant Heating and Cooling Systems
13.3 Thermal Storage Systems
13.4 The Ground as Heat Source and Sink
13.5 Occupant Controlled Windows with HVAC
13.6 Room Air Distribution Systems
13.7 Decoupled, or Dual Path, and Dedicated Outdoor Air Systems
Summary
Your Next Step
Bibliography

## Study Objectives of Chapter 13

Chapter 13 introduces a diverse group of subjects dealing with HVAC. When you have completed the chapter you should be able to:

- State two reasons for using thermal storage.
- Identify two good features of radiant heating and name three examples of where it can be an excellent system choice.
- Describe at least three room air-distribution systems.
- Explain why it can be advantageous to have a separate outside air unit as well as the main air-handler.
- Explain the challenges of having **operable windows**, windows that people can open and close, with an HVAC system.

## 13.1 Introduction

This final chapter covers some special heating, cooling and ventilation applications. We start with radiant heating and cooling, an idea that was partially introduced when we discussed radiant floors in Chapter 8.

From radiant heating and cooling we move on to **thermal storage**. Thermal storage is a method of reducing the need for large equipment and reducing energy expenses. Thermal storage is achieved by having the heating or cooling equipment operate during low load periods, to charge a thermal storage system for later peak-load use. Under certain circumstances, storage of heating or cooling capacity can reduce both installation costs and operating expenses.

From thermal storage systems, we move on to consider the ground as a vast heat source or sink. Following these three sections, we continue with sections dealing with ventilation. The first ventilation topic is a detailed discussion of the issues dealing with operable, 'occupant controlled', windows and the HVAC systems serving these spaces. When occupants are in control of opening and closing windows, there is a largely uncontrolled movement of air in a space. In comparison, following this discussion, we examine the issues of air distribution in rooms that don't have operable windows.

We will discuss various standard ways of delivering air to rooms and their relative merits and popularity. Then, we will take a brief look at separate dedicated outside air units that are particularly valuable in dealing with locations where there is high humidity and substantial outdoor air requirements.

Then it is time to wrap-up with some suggestions for your future.

## 13.2 Radiant Heating and Cooling Systems

As you recall from Chapter 3, radiant heat passes in straight lines from a hotter to a cooler body with no affect on the intervening air.

Radiant heaters and coolers are defined as units that achieve more than 50% of their cooling or heating output through radiation (as compared to convection and conduction). We have already discussed radiant floors and ceilings under the heading '**Panel Heating and Cooling**' in Chapter 8. These panel units operate well below 300°F, and are classified as 'low temperature'. Radiant floors operate at a relatively low temperature, with a maximum surface temperature, for comfort conditioning, of 84°F.

In this section, we will consider high temperature units that operate at over 300°F, revisit radiant floors and briefly consider radiant ceiling panels.

### High Temperature Radiant Units

High temperature, or infrared, units operate at over 300°F. Examples range from units with a hot pipe, to ceramic grids heated to red/white heat by a gas flame, up to electric lamps. These are heaters that are far too hot to get really close to or to touch. There are three main types of high temperature units: high, medium and low intensity.

- High intensity units are electric lamps operating from 1800–5000°F.
- Medium intensity units operate in the 1200–1800°F range and are either metal-sheathed electric units or a ceramic matrix heated by a gas burner.
- Low intensity units are gas-fired, using the flue as the radiating element—basically a gas burner with a flue pipe (chimney) typically 20–30 feet long, with a 4-inch diameter, as shown in *Figure 13.1*. A low-intensity unit operates as a flue that runs horizontally through the space. It will usually, but not always, vent outside and have a reflector over the flue to reflect the radiant heat downward.

These low intensity units can run up to 1200°F, have a dull red glow, and take only three, or four, minutes to reach operating temperature. Since they are gas fired, adequate combustion-air must be provided, as required by local codes.

A single burner low intensity unit is shown in *Figure 13.1*. The blower assembly provides the required forced draft through the burner and long flue. The flue gas temperature drops as it gives up heat along the tube. As a result, the output drops along the length of the unit. Manufacturers can use different strategies to offset this drop in output; tube materials with a lower radiant output in the early sections, or larger tubes in the latter sections.

These strategies are not enough in larger installations, and so units with multiple burners are used. However, multiple-burner units introduce additional complexity into the system. For example, the same forced-draft method cannot be used, since, if one of the blowers failed, the others would blow fumes into the building through the inoperative blower. To avoid this possibility, multi-burner units have an exhaust fan, called a vacuum pump, to draw all the products of combustion from the flue. This is done to ensure that all flue gases are exhausted. This type of arrangement is shown in *Figure 13.2*. For further control of output, high-low, or modulating, burners can be used.

The burner controls are in the self-contained blower-burner assembly, with the whole unit controlled by a long-cycle (slow to respond) thermostat or a proprietary temperature control system. The location of this control is significant.

It is important to remember, from Chapter 3, that both ambient air temperature and the radiant effect of the heater(s) will affect the thermostat. Let us go back to the radiant floor for a simple example. If the thermostat is located on an inner wall (far away from the window), the floor and adjacent warm walls will predominantly influence it. As a result the room tends to be cool for occupants in cold weather, since the cool external wall and windows do not adequately influence the thermostat. This effect is significantly reduced if the thermostat is placed on a side wall (nearer the window), well away from the inner wall so that the cool outside wall and window will have a more significant effect on the thermostat. This alternative could result in the room becoming uncomfortably warm.

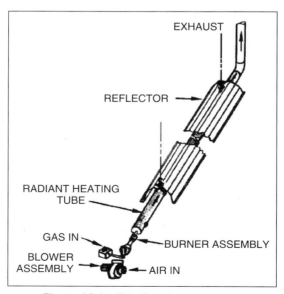

**Figure 13.1** Tube Type Radiant Heater

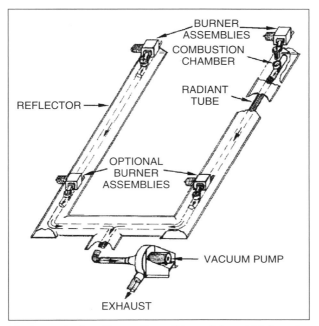

**Figure 13.2** Multi-Burner Radiant Heater (Part of Figure 1, Page 15.2, from Systems Handbook)

The effect of location is even more pronounced with radiant heaters. As a result, it takes skill and experience to make an effective choice of thermostat location. This is one of those occasions when asking, and taking, the advice of an experienced manufacturer can be really worthwhile.

Since the multi-burner radiant-heater units run very hot, they must be out of the reach of occupants. They must also be mounted so that they cannot overheat objects immediately beneath them.

For instance, suppose a machine shop was fitted out with radiant heaters that were mounted 15 feet above the floor. This would provide a comfortable work environment for the staff. However, consider what would happen if the heaters were mounted directly above a floor space that was also used by delivery vehicles that drive into the shop to be loaded or unloaded. In that case, the top of the vehicles would be dangerously close to the heater and could end up with a burnt roof. This problem can often be avoided by designing the heater layout so the heaters are above the work areas only, at a safe distance from vehicle access routes.

Radiant heaters are particularly suitable where high spaces must be heated without obstructing the space, as in aircraft hangers, factories, warehouses, and gymnasiums. They are also valuable where the staff and floor is to be kept warm, but not the space, such as in loading docks, outdoor entrances, and swimming pools.

Radiant heaters are also suitable for racetrack stands and spectator seating around ice rinks. In the ice rink they have the ability to be directed at the seating with a fairly sharp cutoff to prevent heating the ice surface, and they do little to raise the air temperature that would also affect the ice.

### Radiant Cooling

Radiant cooling was introduced in Section 3 of Chapter 8. Radiant cooling is always achieved by using a 'large area' panel system, since the transfer per square foot is quite limited. This is largely because the chilled-water temperature must be kept warm enough to avoid any condensation. The ceiling may be either a plastered ceiling with embedded pipes or a metal pan ceiling with the pipes attached to the panels.

Just like the radiant floor, the radiant cooling ceiling requires no equipment or floor space within the occupied area. With the plaster ceiling there is nothing in the room. This makes it an attractive choice in some hospital situations where cleaning needs to be minimized. Only ventilation air has to be moved around the building and supplied to each room. However, it is critical that the moisture level, relative humidity, in the building be kept low enough to prevent problems that may occur due to condensation on any part of the ceiling panels.

The performance of radiant ceilings is well understood by the various manufacturers of the many different designs and they, and the architect, should be involved early in the design stages. If a metal panel system is chosen, it must fit in with the dimensional requirements of the ceiling. Panels radiate upwards as well as downwards. An un-insulated panel will cool the space below as well as the floor or roof above it. If cooling is not desirable above the panel, the panel can have insulation placed on the top of it. Conversely, if the cooling **is** designed to radiate upwards, be sure that an acoustic pad is **not** specified above a panel, since the acoustic pad will also provide thermal insulation.

One negative of this system is the extended length of time it takes to return the space to comfort levels after the temperature has drifted up. Operators of radiant cooling panel systems need to be aware of the relatively slow response of these systems—even those with light metal panels. As a result, it is not a good idea to allow the temperature to drift up when the space is unoccupied, even though this strategy may *appear* to result in energy savings.

## 13.3 Thermal Storage Systems

Thermal storage systems normally involve the generation of cooling or heating, or both, at off hours while storing this energy for use at a later time, generally to be discharged during peak energy use periods such that overall energy costs are reduced. These systems can be "**active**" or "**passive**".

### 13.3.1 Passive Thermal Storage

"**Passive Thermal Storage**" refers to using some part of the building mass, or contents, to store heating or to store cooling capacity. The very simplest form of passive storage is the choice to construct a building using heavy construction; block walls, block partitions, concrete floors, and concrete roof decks.

During the cooling season, the mass of the building walls and roof can be cooled at night by the air conditioning system, and when favorable, by the cool night air. When the night air is sufficiently cool, then ventilating the building, by either opening the windows or running the ventilation system, can cool the structure. Then, during the day, the sun has to heat the mass of the structure before the inside temperature rises. In addition, the walls and roof have considerable stored heat when the sun goes down and the warm surfaces of roof and wall re-emit a proportion of heat back to the outside.

The interior mass acts as a thermal flywheel, absorbing heat through the day and re-emitting heat through the evening and night. The result is a lower peak cooling load, hence smaller refrigeration equipment is required. In addition, there is a lower total cooling load, due to the heat stored in the day and zre-emitted outside during the night.

Passive water heating is also very popular in warmer climates. A black plastic water-storage tank on the roof will absorb heat through the day, warming the water. If this solar-warmed water is used for the domestic hot-water supply, to wash basins, and for the cold-water supply, to the showers, then hot water is not needed for hand-washing or cool showers. For a hot shower, the already warmed water must be additionally heated by a conventional water heater. This system has the further advantage of operating at low pressure. The system is very energy-efficient but there is the potential hazard of breeding legionella (see Chapter 4) in the solar-warmed storage tank.

There are many excellent books detailing the variations on solar-heated water storage and using the building to store, or reject, solar heat. One word of caution: the local climate makes a huge difference to the overall effectiveness of a solar heating project. For instance, in a climate where the temperature never drops to freezing, water systems need no protection against freezing. In climates where the temperature does drop to freezing, there are two issues to face: first is the shorter proportion of the year when the system can be used, and second, freeze protection is always more challenging than you would expect, so consult with an expert.

### 13.3.2 Active Thermal Storage

Active thermal storage takes place when a material is specifically cooled or heated, with the object of using the cooling or heating effect at a later time.

Perhaps the simplest example is the electric thermal storage (ETS) heater, called a 'brick' or 'block storage' heater in certain parts of the world. These units are commonly used in residences to provide off-peak electric power for heating. The ETS consists of an insulated metal casing filled with high-density magnetite or magnesite blocks. A central electric heater heats the blocks to a temperature as high as 1400°F during off-peak hours, during the night. The units passively discharge through the day and may have a fan to boost output when needed—particularly in late afternoon towards the end of their discharge period.

The units are relatively inexpensive, and, with suitable electrical rate incentives, ETS provide an effective way for a utility to move residential electric heating loads from the day to the night. This allows the utility to level their load, which is almost always to the utility's benefit. This benefit also lowers the energy cost for the consumer, a true win-win situation.

Since the issue of electrical rate structures has been introduced, this is perhaps a good moment to review some of their more typical features.

## Electricity Rate Structures

Virtually all electric-utilities must have users for the power they produce **at the moment** they produce it. Unlike gas, electricity cannot be stored for later use. Electricity has its highest demand period during the weekdays and, in air-conditioned climates, primarily in the afternoon. In order to serve the peak, the utility must have that installed capacity available. That peak capacity sits idle the rest of the day, earning no revenue.

The following description is of a basic electrical rate structure, though there are many other features applied to encourage a balance between the particular utility and their users.

## Consumption Charge and Demand Charge

To balance their costs and income, utilities use two methods of charging those with high peaks in their load. The high peaks are addressed by a "demand charge". The demand charge is, typically, based on the highest load in any 5–15 minute period in the month. The utility meter is continually checking the average load over the previous few minutes and recording the highest peak demand. In addition to the demand charge the utility charges a consumption charge based on the quantity of electricity used. This consumption charge covers all the costs of production.

For example, each month, a utility charges for electricity based on two factors:

**Demand Charge**: $10 per kW of demand (kilowatt = 1,000 watts, equivalent to 10 100-watt light bulbs)
**Consumption Charge**: $0.07 per kWh. (a kilowatt hour, kWh, is the energy used by a 1 kilowatt load in one hour.

Consider a one-kilowatt load on for one hour in a month. It will cost

| | |
|---|---|
| Demand Charge | $10.00 |
| Consumption Charge one hour * $0.07 | $0.07 |
| | $10.07 |

The same heater, on for the whole month (30 days of 24 hours) will cost

| | |
|---|---|
| Demand charge | $10.00 |
| Consumption Charge 30days * 24hours * $0.07 = | $50.40 |
| | $60.40 |

The effective cost of just one hour of operation in the month

$10.07/1 = $10.07 per hour.

The hourly cost for the whole month was

$60.40 / (30*24) = $0.084 per hour.

This is significant encouragement to avoid short peaks!

Large peaks are easily produced with larger chillers. On one campus, the maintenance staff decided to test run two 1,000-ton chillers on a weekday in early spring, the last day of the month. They wanted to make sure the chillers would be ready when the weather warmed up. Adding the two chillers' demand charge for the test run cost over $21,000, simply because the chillers pushed the peak demand up for the month!

**Time-of-use Rate Schedule**

Next, the utility may have a "time-of-use" rate schedule. Earlier we mentioned that low rates encourage the use of night-storage heating through the use of electric-storage heaters. On the other hand, many utilities will charge a hefty premium for power between, for example, noon and 5 p.m. Here the utility is aiming to discourage use in this specific time period in order to minimize their peak.

Both peak demand and time-of-use pricing structures favor the use of thermal storage. In addition, many utilities will give substantial financial incentives to designs that reduce peak demand on their systems. It is always worth checking on what is available and whether the utility will provide financial incentives to help with design in order to maximize savings.

### 13.3.3 Chilled Water and Ice Storage Systems Introduction

Now we are going to move on from passive storage systems and our discussion on electricity rates to consider water and ice storage systems. Why go to the extra effort to use storage? There are two common reasons: to reduce installation costs where possible and to reduce operating costs. Storage is also being increasingly used as emergency cooling capacity for critical installations, such as computer data centers.

**1. To reduce installation costs**:
Consider a specialized-use building, like a church, that has a cooling system designed for the capacity based on the peak attendance that occurs one day a week. For the remainder of the week, though, small attendance is the norm. A small cooling plant and storage system may be much less costly to install and, generally, less costly in electricity bills.

Consider *Figure 13.3*. The chiller is shown running continuously producing almost ten units of cooling capacity. The solid line is the load on a particular day. Starting at the left, midnight, the chiller is serving the load—about 2 units—and the spare capacity is charging the storage. At about 13:00, the load equals chiller capacity and from then until 21:00, the load over-and-above chiller capacity is met from storage. Effectively, the excess chiller capacity at night has been stored for use during the high load in the afternoon.

In some situations this lower installation cost may be achieved even with full daily usage. Factors that can contribute include: smaller chiller, smaller electrical supply, a financial incentive from the utility, and, when ice is the storage medium, even smaller pumps, pipes, fans and ducts are possible.

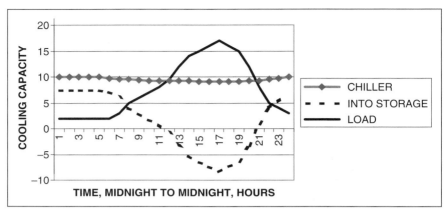

**Figure 13.3** 24 Hour Cooling Load Profile

**2. To reduce operating costs**:
We have already discussed demand and time-of-day pricing structures that encourage night-time use and discourage afternoon use. As demonstrated, it can be worthwhile to run the chiller during the night and on weekends to avoid demand charges, and overnight and in the morning to avoid time-of-use charges.

## Chilled Water Storage

Let's consider water first. Water holds 1 Btu/lb for every 1°F change in temperature. If our stored water is available at 41°F and return temperature from the cooling coils is 56°F, then every pound will have a storage capacity of 56 Btu − 41 Btu = 15 Btu. A cubic foot of water weighs 62.4 lbs, so a cubic foot of our stored water represents:

15 Btu/lb · 62.4 lb/ft³ = 936 Btu/ft³

By definition, a ton of air conditioning is 12,000 Btu/hr, so, theoretically, to store a ton-hour will require:

12,000 Btu/ (936 Btu/ft³) = 12.8 ft³

In fact, it will require 10% to 50% more, since there are the inevitable losses in the system as the water is pumped in and out, as well as heat gains through the insulated tank wall.

Chilled-water storage is generally conducted with normal, or slightly lower than normal, chilled-water supply temperatures. As a result, producing chilled water for storage can be done using a standard chiller running at approximately the same efficiencies used for conventional chilled-water systems. Chilled-water storage systems tend to dominate the large-system market with tanks that have capacities of half a million cubic feet and more.

Now, let's consider the use of ice for thermal storage. One cannot make and store a solid block of ice; one needs a mechanism to get the heat in and out. For the sake of example, let's assume 70% of our storage volume is ice and our system simply recovers the "**latent heat of fusion**". The latent heat of fusion of water is 144Btu/lb, which is the heat absorbed to melt one pound of ice or convert one pound of water to ice at 32°F. The latent heat of fusion of 1ft³ of ice is

144 Btu/lb · 62.4 lb/ft³ = 8986Btu/ft³.

In our example, only 70% of the volume can be ice, so the latent heat of fusion storage would be

8,986 Btu/ft³ · 0.7 = 6290 Btu/ft³.

This means chilled water requires about four to seven times the storage volume that ice requires for the same amount of cool storage volume. So, the big advantage of using ice storage is that a much smaller volume of storage is required. However, to achieve this small volume, the chiller must produce much lower discharge temperatures, below 30°F, instead of 40°F, so the chiller efficiency is lower. In addition, the production and handling of an ice storage system generally requires a more sophisticated plant. This smaller space requirement makes ice storage generally more popular for single buildings.

As a result, (to be very simplistic) there is a choice between:

1. Water: A relatively simple and more efficient chilled-water production with larger storage-space requirements.
   *or*
2. Ice: A relatively more complex system with a less efficient chiller, producing ice and using a smaller storage space requirements.

These underused techniques of water and ice storage are clearly explained in considerable detail in ASHRAE's *Design Guide for Cool Thermal Storage*.

In the next sections, we will discuss the basics of practical water and ice storage systems.

### 13.3.4 Chilled-water Storage

Storing chilled water is normally done in a large **stratified tank**, cold at the bottom and warmer at the top. Stratification is required to avoid mixing warmer and cooler water while the tank is charged and discharged. Conveniently, water has a maximum density at 39.2°F. So, water that is warmer than 39.2°F will float above water that is at 39.2°F.

Chiller water enters the bottom of the tank, at low velocity, through a diffuser, as shown in *Figure 13.4*. Typically, the diffuser is a loop, or an array of pipes with slots, to allow the water in or out with minimal directional velocity, to minimize mixing. The chilled water enters at 40°F (just above 39.2°F) and the warmer water at the top stays stratified above the 40°F water. As more warm water is pumped from the top of the tank, through the chiller, and returned very gently to the bottom of the tank, the cold layer gradually moves up the tank. When discharging, chilled water is withdrawn at the bottom of the tank and an equal volume of warmed water is returned to the top of the tank. A similar diffuser at the top of the tank minimizes turbulent motion and mixing in the water. The process produces a thermal gradient in the tank, such as shown on the right of *Figure 13.4*.

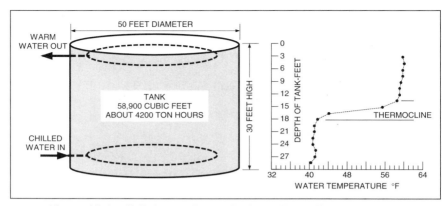

**Figure 13.4** Chilled-water Storage Tank with Typical Thermal Gradient

In *Figure 13.5*, a simple circuit is shown with the loads and chiller-circuits below the water level of the storage tank. Valves to control the flows between tank and chiller are not shown in *Figure 13.5*, since there are several alternatives. There are two pipe-loops connected to the storage tank: one belongs to the

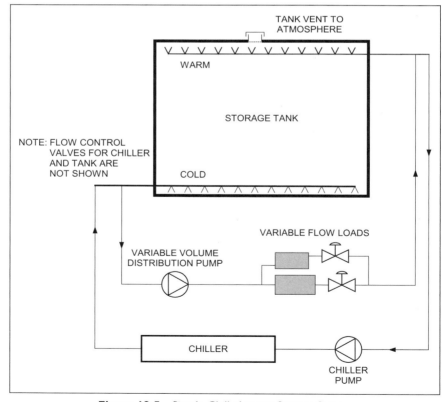

**Figure 13.5** Simple Chilled-water Storage System

| STORAGE | Charging | Charging | | Discharging | Discharging |
|---|---|---|---|---|---|
| CHILLERS | Charging | Meeting load | Meeting load | Meeting load | Off |
| LOADS | Off | On | On | On | On |

**Figure 13.6** Possible Storage Operating Modes

chiller and the chiller pump, and the other is the load circuit, consisting of the variable volume pump, and variable flow loads.

There are up to six possible operating conditions with a storage system, as shown in *Figure 13.6*.

To maximize savings, the designer must give special consideration to the control of larger storage systems. The seasons when full tank capacity is not required are a particular challenge. On one hand, it is wasteful to over store. On the other hand, if you under store, then you could be faced with much higher electricity charges, or a lack of sufficient capacity at peak load periods. Because these penalties are usually much more costly than any savings that could be achieved by reducing storage, full storage is generally used.

For maximum storage, the temperature difference between the chilled-water supply and return water must be as large as possible. In general, chilled-water storage is not economical with a temperature differential below 12°F. A storage temperature difference of 20°F should be the target to make the system as economical as possible..

Chilled-water storage is not high-tech. Water tanks are a common item in both steel and concrete and the controls do not have to be very complex. Chiller efficiencies are, often, lower because of the lower chilled-water supply temperature required. (Remember from Chapter 6, Section 6–3, the efficiency of a refrigeration circuit falls as the difference in temperature between evaporator and condenser increases.) However, the chiller efficiency that can be achieved is maximized, since the chiller can always run at full load and the operation is largely at night when ambient temperatures around the cooling towers are lower, allowing a lower condenser-water supply-temperature. Efficiency can also be improved by using a larger cooling tower, which will drop the condenser-water supply-temperature

Exposed tanks should be insulated to minimize heat gain to the cooled stored water. The size of the storage tank should allow for:

heat transfer and mixing between warm and cold water levels
ambient heat gain
pumping power

The net useful cooling output typically varies between 80% and 90% of the input cooling.

One particularly effective use of chilled-water storage is in the capacity extension of existing facilities.

For example: suppose the client has a building that is running well and needs a substantial addition. You could choose to buy additional chiller capacity for the additional load. Alternatively, it may be far more economical, on space, installation cost and operating cost, to add chilled-water storage and have the existing plant run more continuously through the evening, to serve the increased load.

## 13.3.5 Ice Storage

There are four main methods of generating ice for ice storage systems: coils, with external melt; coils, with internal melt; ice harvesting; and water in numerous plastic containers.

In **External melt** systems, ice forms around coil of pipe in a tank. The coils are cooled, and may be steel or plastic. Just two of the pipes in the coil are shown in *Figure 13.7*. The pipes are spaced so that when fully charged with, for example, 2.5 inches of radial ice, there is still space for chilled water to flow between the iced pipes.

In **Internal melt** systems the pipes are closer together and cold brine—water containing an antifreeze chemical—passes through the pipes, which causes a block of ice to form around the pipes. To discharge, warm brine passes through the pipes melting the ice around them.

**Ice harvesting** systems generally have a set of vertical flat, hollow panels above a tank of water, as indicated in the schematic, *Figure 13.8*. The panels cycle between two functions, first as a chiller evaporator, and then as a condenser, just like the heat pump circuit we discussed in Chapter 6.

The process begins with the panels acting as the chiller evaporator: Water, is continuously pumped over the plates and a layer of ice begins to form on the plates. After 20–30 minutes the ice reaches an optimum thickness and the refrigerant cycle is reversed. Then the panels act as the condenser: The hot condenser gas then melts the ice at the plate surface and it falls into the tank.

Ice harvesting systems are attractive since they can be purchased as factory designed-and-built systems. If needed, they can have a very high discharge rate, and the full 24-hour charge can be removed in as little as half an hour.

Cooling is removed by passing return chilled water through the ice harvester and ice-water storage tank to achieve a chilled-water supply temperature of 34–36°F. This is much colder than the 42°F, or warmer, water from standard chilled-water systems, even those using chilled-water storage.

Lastly, water can be contained in plastic spheres. The spheres are either partially filled with water with some air to allow for expansion on freezing or the spheres have depressions which fill out as the water expands, as it freezes.

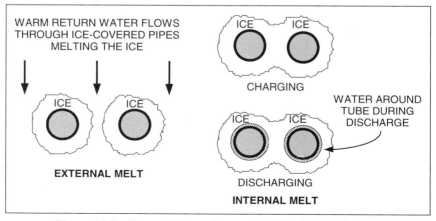

**Figure 13.7** External Melt and Internal Melt Ice Storage Systems

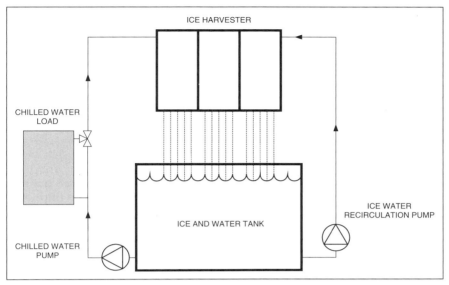

**Figure 13.8**   Ice Harvesting

In these systems, chilled water containing an antifreeze flows through a tank full of these spheres, to either store or extract cooling.

The major advantages of ice-storage systems are smaller storage tanks and lower chilled-water supply temperatures. The lower chilled-water supply temperatures can be used to increase the system water-temperature differentials and to produce very cool, low temperature, supply-air for distribution to the building's occupied spaces. This results in smaller pipes and smaller air-distribution ducts and supply-air fans. The low-temperature air-supply system does require carefully designed diffusers that do not dump cold air onto the occupants.

The cooler chilled-water supply temperature from ice storage can be very useful in extending an existing chilled-water system. Suppose there are several buildings on a main chilled-water loop and the client wants to add another building at the end, farthest from the chilled-water plant. The option of increasing the chilled-water pipe-size may be prohibitively costly and disruptive. By adding ice-storage, the chilled-water supply-temperature can be reduced from 42°F to 35°F. If the original system was designed for 42°F chilled-water supply and 55°F return, the temperature rise was

55°F − 42°F = 13°F.

Now with chilled water at 35°F the temperature difference is

55°F − 35°F = 20°F.

With the same volume flow, the capacity of the piping mains has been increased by 50%, which now allows this system's pipes to serve the remote building without replacing them with larger pipes. Adding insulation to the existing pipes may be needed.

204    Fundamentals of HVAC

To achieve the projected savings in energy costs, if the system is not fully automated, the operating staff must completely understand and be able to apply the control strategies of the design. With today's technology, these can be performed by Direct Digital Controls (DDC), through software. Using DDC, the control sequences can be made fully automatic and therefore, less dependent on the operating staff. However, this does require that these systems be commissioned to ensure that the automatic control functions as intended.

Be warned that it is surprising how often operating and maintenance staff defeat the cleverest software by switching just one piece of equipment to 'manual'!

There are several other, less popular, active storage methods that you can research elsewhere. Be aware that 'less popular' does not mean 'unpopular'. Many systems are ideal choices for some specific situations but are not practical for every project. Local knowledge and your research can help find the best choice for your project.

## 13.4 The Ground as Heat Source and Sink

The ground can be treated as a large heat source or as a heat sink. In other words, one can extract heat from the ground or reject heat to the ground. The temperature only a few feet below the surface varies half as much as the ambient temperature. Below 10 feet the temperature remains fairly constant in most places.

There are three general methods of using the ground as a heat source or sink: the well, the horizontal field and the vertical field.

**The Well**: The oldest method, and, in some places, the easiest, is to dig a well, then pump the water up and through the heat pump before piping it to drain. Many local codes will not permit this approach and will require you to have a second well some distance away to discharge the water back into the ground. This all assumes your location has a readily accessible, adequate and reliable flow of 'sweet' water. 'Sweet' meaning it has no undesirable characteristics, such as dissolved salts that will corrode away both pumps and heat exchangers very quickly. Local knowledge and test holes can be invaluable.

The horizontal field and vertical field refer to pipe loops in the ground that transfer heat to or from the ground.

**The Vertical Field**:

1. The field has been prepared and planned, and then vertical bore holes are drilled.
   The vertical depth for the boreholes ranges from 50 to 500 feet, depending on ground conditions and the cost to drill the holes to these depths. Boreholes must be spaced well apart to avoid having them thermally affecting each other. The effect is minimized with a row of holes, but this is not always an attractive alternative. A rule of thumb is 20 feet apart, but local conditions, such as underground water flow, can reduce this distance. A test hole can be bored and used to test the heat transfer characteristics of the local soil conditions to help determine the number of wells and spacing required.
2. Durable U-shaped plastic pipe loops are lowered into the boreholes.

3. Each borehole is back-filled with excavated material or with a special mixture to enhance heat transfer with the ground.
4. The ends of the pipes are connected to headers, which are routed back to a building to pumps within the building. The pumps are connected to piping that is circuited to one, or more, water coils, each on one side of a heat pump.

Vertical ground source systems have the following advantages:

- They utilize smaller areas of land than the horizontal system.
- Their performance is quite stable(when spaced and sized properly), since the ground temperature does not vary with the seasons.
- They use the lowest pumping energy and the least amount of pipe.
- They often provide the most efficient performance.

Vertical ground source systems have two disadvantages that vary according to location:

- They are generally more expensive to install than horizontal systems and can be prohibitively expensive in hard rock areas.
- The availability of qualified contractors is very limited in some areas.

**The Horizontal Field**: This method involves burying pipe loops in trenches or open pits at a depth of at least 4 feet. There is a variety of pipe loop arrangements that are designed to take advantage of local conditions.

Horizontal systems have the following advantages:

- They are relatively easy to install with readily available, non-specialist, equipment in areas without rock.
- For rural residential systems, the land requirement is usually not a restriction.
- They usually have a lower installed cost than vertical systems and they are potentially easier to repair.

Disadvantages

- They require a much larger land area.
- They have a more significant variable system performance than the vertical arrangement, due to greater variations in ground temperature that arise from seasonal temperatures, rainfall and shallower burial depth.
- Their efficiency is generally lower than the vertical arrangement, due to fluid temperature and slightly higher pumping requirements.

Correctly sizing a heat pump for winter heating and summer cooling can be a difficult task. In many climates with cold winters, the winter heating load can be much higher than the summer cooling load. Installing a heat pump that is big enough to do both tasks is often a mistake. If the unit is oversized for summer cooling, it will cycle excessively and dehumidification will be very poor to non-existent. The maximum over-sizing above summer load should not exceed 25% for reasonable summer performance. The winter heating load that is not supplied by the heat pump is best provided by supplemental heat.

One relatively new opportunity to deal with this issue is the two-speed compressor unit. Two-speed units may allow for correct sizing for the summer load by cooling at low speed, while high speed may allow the winter heating load to be more closely met. For heating in these climates, it is very efficient to use a radiant floor system. This is because the temperature difference between the ground and the heat pump heating-supply temperature is lower, thereby, providing a significantly higher efficiency.

An extension of the idea of using "natural" sources for heating or cooling is the idea of using natural ventilation from operable windows. This will be covered in the next section.

## 13.5 Occupant Controlled Windows with HVAC

People like to think they have control of their environment. For air-conditioned buildings without operable windows, there is a desire "to have a thermostat in my office." In fact, many maintenance staff have discovered that the presence of a thermostat can be very satisfying even when it is not connected! Hence the use of the phrase 'dummy thermostat.'

This desire for control is often successfully exercised in the demand for occupant controlled windows, operable windows. Unfortunately, people are not good at assessing when to have the window open or when to close it. This is where good communication can have a very beneficial effect. People are generally cooperative if they understand why they should be cooperative. You would be surprised at how many buildings have occupants running window air-conditioners while the windows are open. The owners make no effort to explain the waste and lack of dehumidification that occurs when the air-conditioner is cooling while the window is open on a hot, humid day. The result is fewer satisfied occupants and the owner has a higher electricity, or energy, bill. If you are faced with a situation like this, try to let the occupants know the benefits that will affect them if they use the system more efficiently.

Actual ventilation depends on orientation, building height, wind direction and wind speed. In narrow buildings with windows on both sides, a cross flow can be very effective. One problem is that on the incoming side occupants may experience an unacceptable draft if they are close to the windows.

In winter, in colder climates, the warm, less dense air in buildings tends to rise. As a result, there is a constant inflow of air through openings that are low in the building and a outflow high in the building. An occupant who opens a ground floor window in a three story apartment building receives an incoming icy blast. The window is quickly shut and remains closed. On the other hand, the person on the third floor can open their window wide and the warm air from the building will flow outward. They can leave their window open, letting the warm air, and energy, of the building continuously vent outside. In this situation, the windows are unusable low in the building and a great waste of energy for negligible ventilation high in the building. The problem of providing enough ventilation without a huge energy waste is addressed in Canada and parts of northern Europe by requiring mechanical ventilation in residences. This has, in turn, made a variety of heat recovery units quite popular, and in many places mandatory, although their cost is often not recovered from the energy savings when fan power is included in the calculations.

In mild climates, operable windows can be used to both ventilate the building and provide overnight pre-cooling with judicious building design and use.

## 13.6 Room Air Distribution Systems

In **mixing** ventilating systems, the air is supplied, typically at 55–57°F, at a velocity of over 100fpm, (feet per minute), through an outlet diffuser or grill, at the ceiling or high in the sidewall. The objective is to have the supply air entrain and circulate the room air, to achieve good mixing. *Figure 13.9*.

The flow from a typical ceiling diffuser has a velocity profile as shown in *Figure 13.9*. The air velocity falls as more room air is entrained and the design should have the velocity no higher than 50 fpm in the occupied zone. When cooling, as shown on the left in *Figure 13.7*, the cool air is blown out across the ceiling and, although cool and dense, does not immediately drop due to the "Coanda" effect. The **Coanda effect** is the property of air to stay against a surface. For the cool air to drop from the ceiling, room air would have to move in above it, since otherwise a vacuum would be formed. This takes time to occur, with the result that the cool supply air travels far further across the ceiling before dropping than would the same flow if it had been discharged well below the ceiling.

The ceiling diffuser works well in the cooling mode. Unfortunately, it does not work very well in heating mode, since the warm, less dense, supply air stays up at the ceiling, out of the occupied zone. The buoyancy effect is particularly problematic with the supply air temperature more than 15°F higher than the general room temperature. The flow is shown on the right of *Figure 13.9*. The air enters the room and stays at the ceiling level except where the cool window creates a downdraft that provides a cool to cold draft over the occupants' feet.

Mixing works well for cooling and can produce an even temperature throughout the space. Disadvantages include:

- The air velocity has to be low enough throughout the occupied area to avoid drafts, so there is a tendency for inadequate air movement in some areas.
- Any pollutants in the space can be spread throughout the space.
- All loads must be absorbed within the mixed air.

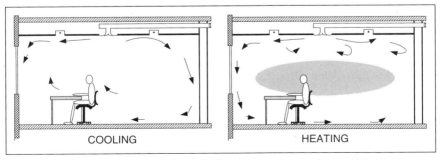

**Figure 13.9** Ceiling Diffuser Airflow Pattern for Cooling and Heating

**Displacement** ventilation is the opposite of mixing. Displacement ventilation aims to avoid mixing in the occupied zone. Air, a little cooler than the space, is introduced at a low velocity (<100 fpm) through large area diffusers in the wall close to the floor. The air flows slowly and steadily across the space until it passes a warm object–a person or a piece of equipment. The warmth causes some of the air to rise up out of the occupied zone carrying pollutants and heat with it. Above the occupied zone, mixing occurs and the return outlet at the ceiling level draws the some of the mixed air out of the space. The flow pattern is shown in *Figure 13.10*.

The air supplied cannot be more than about 7°F less than the occupied space temperature, in order to avoid excessive cooling on the people closest to the outlets. This restriction severely limits the effective cooling capacity of the system. For cooler climates, such as Scandinavia, where the system is very popular, this load restriction is not as significant. Where higher internal loads must be absorbed, there are methods of entraining room air into the supply air to increase the effective flow into the room while still staying within 7°F less than room temperature.

The air movement in the space separates into the lower displacement zone with a recirculation zone above. In a well-designed space, the recirculation zone is just above the occupied zone.

The objective of the system is to have the occupants and the equipment in a flow of clean air, with their own heat causing convection around them. This will lift their pollutants up, out of the occupied zone. In addition, the convection heat from surfaces and lights above the occupied zone do not affect the temperature in the occupied zone. As a result the air leaving the room can be warmer than would be acceptable in the occupied zone.

**Under Floor Air Distribution**, UFAD, is supplied from a raised floor through numerous small floor grilles. The floor typically consists of 24 inch square metal plates, or tiles, supported by a 10–18 inch high supporting leg, or column, at each corner. Some of the tiles have outlet grilles installed in them. The tiles can be lifted and moved around, making grille re-location, addition, or removal, a simple task as shown in *Figure 13.11*.

Air, at 58–64°F, is supplied to the cavity and discharges through the floor grilles. The floor grilles are designed to create mixing, so that the velocity is below 50 fpm within 4 feet of the floor. You can think of the air as turbulent

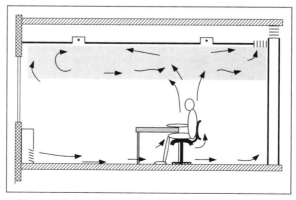

**Figure 13.10** Schematic of Displacement Ventilation

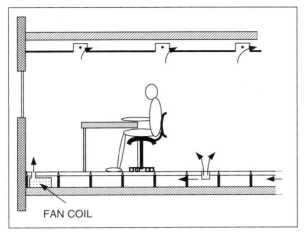

**Figure 13.11** Under-Floor Air Distribution (UFAD)

columns spreading out above the 4-foot level to form a vertical displacement flow towards the ceiling. Return air is taken from the ceiling or high on the wall. The rising column of air takes contaminants with it up and out of the breathing zone. This sweep-away action is considered more effective rather than mix-and-dilute. As a result, the ventilation requirements of ASHRAE Standard 62.1 can be satisfied with 10% less outside air.

There are numerous outlets, since the individual outlet volume is typically limited to 100 cfm. The entering air does not sweep past the occupants, as occurs in displacement ventilation, so there is no restriction on cooling capacity. There is, however, a limit on how well the system will work with rapidly changing loads. For spaces with high solar loads, thermostatically controlled fans or other methods are required to modulate the capacity to match the changing load.

Since the air is rising towards the ceiling, the convection heat loads above the occupied zone do not influence the occupied zone temperature. Therefore, the return air temperature can be warmer than the occupied zone and a return air temperature sensor is a poor indicator of occupied zone temperature.

The cool air flows continuously over the structural floor that somewhat acts as a passive thermal storage unit. This storage can be used to reduce peak loads.

For perimeter heating, small fan-coil units can be installed under the floor, using finned hot water pipes or electric coils. In a similar way, conference rooms that have a highly variable load can have a thermostatically controlled fan to boost the flow into the room when it is in use.

A modification of the under-floor system with individual grilles is the use of a porous floor. The floor tiles are perforated with an array of small holes, and a porous carpet tile allows air to flow upwards over the entire tile area. This is a modification of the standard grill and has yet to gain popularity.

The under-floor air delivery system has the following advantages:

- Changing the layouts of partitions, electrical and communications cables is easy. For buildings with high "churn" (frequent layout changes) this flexibility may, in itself, make the added cost of the floor economically justified.

- The flow of air across the concrete structural floor provides passive thermal storage.
- When the main supply duct and branches to the floor plenums are part of a well-integrated architectural design, the air supply pressure drop can be very low, resulting in fan-horsepower savings.
- Less ventilation outside air can potentially be used.

Disadvantages include:

- A significant cost per square foot for the floor system supply, installation and maintenance.
- A tendency to require a greater floor-to-floor height, since space for lights and return air ducts is still required at the ceiling level.

Our fourth and final type of air distribution system is most often a variation of the under-floor system. It is the **Task\Ambient Conditioning system, TAC**. With TAC each occupant workstation is supplied with cooling air and a degree of control over this airflow, airflow direction and temperature, as shown in *Figure 13.12*. In a typical arrangement one, or two, supply air nozzles are mounted above the work surface. The occupant can easily alter the velocity and direction of flow. Temperature may be controlled by mixing room air into the supply air, or by a resistance or radiant electric heater controlled by the occupant.

The ability to control their own environment is very popular with the occupants, though the measured conditions are not greatly different from occupants in the same building without a TAC. One specific advantage of the TAC for the occupant is the ability to modify the air speed. Since this system is in addition to the under-floor supply, there is significant research work being done to prove that the cost is more than recovered in improved staff productivity.

This completes our look at supplying air to occupied spaces. As with so many issues in HVAC, the climate and the local norms and experience will often drive decisions as much as technical merit.

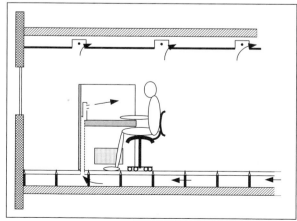

**Figure 13.12** Task/Ambient Conditioning Supplied from Under-Floor Distribution

Special Applications    211

Having discussed room air distribution we are now going to move to the other end of the system, where the ventilation air is brought into the building through the air handler.

## 13.7 Decoupled or Dual Path, and Dedicated Outdoor Air Systems

Our last area of discussion relates to outdoor air. There are situations where mixing the outdoor air with return air and conditioning the mixture is not a good choice.

Let us consider the following example: a humid climate, on a cloudy, very high humidity day that is warm, but not hot.

The typical package air-conditioning system will do a poor job, since the cooling coil will take out very little moisture, because there isn't adequate sensible load to keep the unit running continuously. The challenge is shown on the psychrometric chart, *Figure 13.13*.

> **Point 1** is the outdoor air at 80°F and 80% relative humidity.
> **Point 2** is the return air from the space at 75°F and 55% relative humidity.
> **Point 3** shows 20% outside air (Point 1) mixed with 80% return air from the space (Point 2).
> Let us assume that the cooling load only requires cooling the air to 65°F.
> **Point 4** shows this air cooled to the required 65°F. Unfortunately, the condition of the mixed and cooled air at Point 4 contains more moisture than the space.

If this air were introduced into the space, the relative humidity would rise until some equilibrium balance was achieved. To prevent this uncontrolled

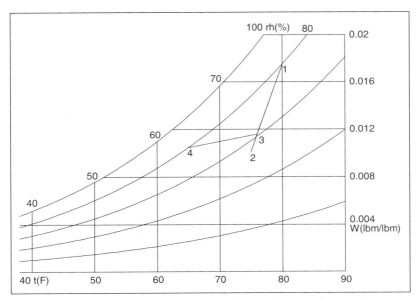

**Figure 13.13**   Ineffective Performance of Cooling Coil for Moisture Removal

increase in moisture, the air going through the coil must be cooled substantially more than is needed for sensible cooling. This is generally not acceptable, as the overcooling would be have to be offset by some form of reheating. Alternative methods of moisture removal are necessary.

This can be achieved in many ways. One way is by treating the outside air before it is introduced into the main air-handling unit. A single cooling coil, designed for the low outdoor air volume and high dehumidification load, may cool and dehumidify this outside air. Typically, this is a deep coil, with a low air-velocity that provides enough time for substantial moisture removal.

In *Figure 13.14*, we see the diagram illustrating this method:

**Point 1** is outside air at the same conditions of 80°F and 80% relative humidity.
**Point 2** is the condition of the return air that is mixed with air from the **new Point 3**.
**Point 3** is air that has been cooled and dehumidified to 55°F and 95% relative humidity—a condition that has a much lower moisture content than the space. (Remember, the higher relative humidity at a lower temperature can still mean a lower moisture content.)
**Point 4** shows that the mixed air has a lower moisture content than the return air from the space.

If the outside air is 20% of the mixture, it provides about 20% sensible cooling, leaving the main cooling coil to do only as much additional sensible cooling as is necessary.

Another method of achieving the required dehumidification is to provide a bypass around the main cooling coil. A part of the air, let us say 50%, flows through the main cooling coil. This 50% flows at half the velocity through

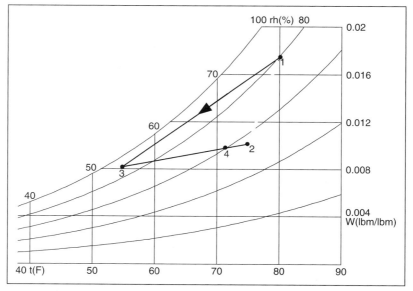

**Figure 13.14** Cooling and Dehumidifying Outside Air Before Mixing with Return Air

the main cooling coil, allowing the air to cool down and condense significant moisture. The other 50% of the air bypasses the coil before mixing with the sub-cooled air. The two air streams then mix to produce a mixture with half the sensible cooling and well over half the latent cooling (moisture removal), much better than if no air bypassed the coil. Another variation of this is to bypass only drier room return-air around the cooling coil and have a portion of the return air mix with the outside air, which is then sub-cooled as it passes through the coil.

We have briefly considered using alternative arrangements to deal with high moisture removal. Now we will consider a situation where different requirements make a dual-path system attractive.

Consider a building that includes a large kitchen and an eating area. The building could be designed to have all the necessary kitchen makeup air come in through the main air handler. However, because the kitchen is a more industrial-type environment, rather than an office-type environment, the kitchen makeup air does not need to be conditioned to the same moisture and temperature conditions as the main air supply to the building. In addition, the kitchen may start operation before the rest of the building and shut down well before the rest of the building. This is a case of a mismatch in requirements and a mismatch in timing.

Therefore, it is often better to provide the kitchen makeup air from two sources. First, there is the air from the eating area. In order to avoid distributing food smells around the building, this air from the eating area should not be returned to the main air handler. Instead, it should form the first part of the kitchen exhaust hood makeup air. The transfer can be by a plain opening, an open door, or a duct with a fire damper, depending on local codes and design requirements. The rest of the kitchen exhaust makeup air can be provided from a separate air handler designed to condition the incoming air to provide suitable kitchen working conditions, often a much less onerous requirement.

## Summary

### 13.2 Radiant Heating and Cooling Systems

Radiant heaters are defined as units that have more than 50% of their heating output achieved through radiation.

**Radiant Heating**: High temperature, or infrared, units operate at over 300°F. There are three main types of high temperature units:

- High intensity units are electric lamps operating from 1800–5000°F.
- Medium intensity units operate in the 1200–1800°F range and are either metal-sheathed electric units or a ceramic matrix heated by a gas burner.
- Low intensity units are gas-fired and use the flue as the radiating element

Important safety and control issues to consider include both heater location and thermostat location.

**Radiant Cooling**: This is always achieved by using a 'large area' panel system. Issues for consideration include: space moisture level, location of insulation on the panels, and the response time of the system.

214    Fundamentals of HVAC

## 13.3 Thermal Storage Systems

Thermal storage can be "**active**" or "**passive**".

*Passive thermal storage* uses some part of the building mass or contents like a thermal flywheel to store heat or cooling and to release it over time to reduce the heating or cooling load.

*Active thermal storage* takes place when a material is specifically cooled or heated, with the object of using the cooling or heating effect at a later time.

### Chilled Water and Ice Storage

There are two reasons to use chilled water and ice storage: to potentially reduce installation costs and to reduce operating costs.

*Chilled-water Storage*: Storing chilled water is normally done in a large stratified tank, cold at the bottom and warmer at the top. One particular economical use of chilled-water storage is in the capacity extension of existing facilities.

*Ice Storage*: There are four main methods of generating ice for ice storage systems: coils with external melt; coils with internal melt; ice harvesting; and water in numerous plastic containers. Ice storage can result in smaller pipes, ductwork, and fans, when low-temperature supply-air is used. Ice storage requires less space than water for the same storage capacity.

## 13.4 The Ground as Heat Source and Sink

The ground can be treated as a large heat source or as a heat sink: one can extract heat from the ground or reject heat to the ground. There are three general methods of using the ground as a source or sink, the well, the horizontal field and the vertical field.

## 13.5 Occupant Controlled Windows with HVAC

People like to think they have control of their environment, resulting in a demand for occupant controlled windows, operable windows. Unfortunately, people are not good at assessing when to have the window open or when to close it.

Actual ventilation depends on orientation, building height, wind direction and wind speed. In mild climates, operable windows can be used to both ventilate the building and provide overnight pre-cooling with judicious building design and use.

## 13.6 Room Air Distribution Systems

There are four main types of room-air distribution: mixing, displacement, under-floor, and task control. Mixing is by far the most popular in North America and task control has yet to gain popularity.

### 13.7 Decoupled, or Dedicated Outdoor Air Systems

There are situations where mixing the outdoor air with return air and conditioning the mixture is not a good choice, like in warm, humid climates; or where fumes should not be recirculated with the building air.

## Your Next Step

The objective of this course has been to provide you with an understanding of HVAC in general, and to introduce you to the more common systems used in the HVAC industry. We have not gone into great detail on any subject but hope to have provided you with enough knowledge to understand how systems work and to decide what you want to know more about.

### Fundamentals Series

For further study, ASHRAE has the following titles in this Fundamentals Series.

- Fundamentals of Thermodynamics and Psychrometrics
- Fundamentals of Heating and Cooling Loads
- Fundamentals of Air System Design
- Fundamentals of Water System Design
- Fundamentals of Heating Systems
- Fundamentals of Electrical Systems and Building Electrical Energy Use
- Fundamentals of HVAC Control Systems
- Fundamentals of Refrigeration

### ASHRAE Handbooks

The four ASHRAE Handbooks are an excellent source of information on all aspects of HVAC and R. One volume is updated and published each year on a four-year cycle. Members receive a copy of the current year's edition each year and copies can be individually purchased.

All four handbooks can also be obtained on a CD.

**Fundamentals** – This volume contains information on the properties and behavior of air, water and other fluids, and how they flow in ducts and pipes. It includes the theory and practice of calculating heat gains and heat losses through all types of building materials.

**Systems and Equipment** – This volume includes HVAC systems, air handling and heating equipment, package equipment, and general components such as pumps, cooling towers, duct construction and fans.

**Applications** – This Handbook begins with a section on how to apply systems and equipment to comfort, industrial and transportation situations. Following this, there is a section on general issues, such as operation and maintenance, and energy management. The Handbook finishes with general applications such as the design of intakes and exhausts, seismic restraint, water treatment and evaporative cooling.

**Refrigeration** – This volume provides very detailed information on all aspects of refrigeration equipment and practices, followed by sections on food storage, food freezing, low temperature refrigeration, and industrial applications that include ice rinks.

The ASHRAE Handbooks are, as a matter of policy, not commercial. They do not recommend any product. Therefore, they lack the reality (or dreams?) of the manufacturers' sales and engineering materials. Don't hesitate to ask manufacturers for sales materials and read them with an alert mind. Is there something here that could really work well in this situation? Is this too good to be true? If so, why? Be realistic, manufacturers put the best light on their product. The challenge for you is to find the product that will perform well in your situation.

### Manufacturers

Manufacturers put significant effort into training their staff about their products. Do not be shy to ask them about their products. When choosing a product, ask the representative: "What would you suggest?" "Is it suitable?" "Is there something better?" "Is there something less expensive?" "Is there something more efficient?" "Who has one of these in and working and can I call them?" Be sure to ask more than one manufacturer's representative for information, so you can get a different perspective on what is available for your application.

Keep asking, keep learning and have fun doing it.

## Bibliography

1. ASHRAE Handbook 2001 Fundamentals
2. ASHRAE Handbook 2002 Refrigeration
3. ASHRAE Handbook 2003 Applications
4. ASHRAE Handbook 2004 Systems and Equipment
5. ASHRAE Design Guide for Cool Thermal Storage, 1993
6. ASHRAE Underfloor Air Distribution Design Guide, 2003
7. ASHRAE Humidity Control Design Guide, 2001

## Epilogue

This story is not a part of the text of the book. I have heard and read a number of variations of it over the years. To me, it speaks of the importance of what we are doing, and what we can be doing, as members of this profession:

*Long ago a king decided to go out on his own to see his kingdom. He borrowed some merchant's clothes and dressed so that no one would recognize him.*

*He came to a large building site and went in while the gatekeeper was dealing with a delivery of huge wooden beams. As he walked around the site he came upon a stonemason, who was chiseling at a large piece of stone.*

*"What are you doing?" the king asked.*

*"Oh, I'm making this stone to fit that corner over there." said the man, pointing.*

*"Very good." Said the king, and walked on. The king approached another stonemason and asked "What are you doing?"*

"I'm doing my job. I'm a stonemason. It's great working here, lots of overtime, enough to pay for an extension to the cottage." said the man with a big grin.

"Very good." said the king, and walked on.

The king stood and watched the third stonemason, who was carefully working on a detail, before asking him "What are you doing?"

The man paused, and looked up, considering his reply. Then he answered "I am building a cathedral."

# Index

acoustical environment 7
active thermal storage 195–7, 214
activity level, and comfort 8, 33
actuator 72
adiabatic process 16, 30
air-and-water systems 25
air-conditioning systems
  basic system 20–3
  choice of system 26–30, 31, 82, 87
  components 20–1, 30–1, 70–5, 86
  controls 26
  definition 4
  economizer cycle 21–3, 31, 89
  processes 3–4, 9
  rooftop units 82–5, 87
  single-zone systems 68–87
  split systems 85, 87
  system performance
    requirements 80–2, 87
  window air-conditioners 4, 26, 76–8,
    107, 177–8
  zoned systems 23–6, 31
air distribution systems 207–11, 214
air handlers *see* single-zone air handlers
air inlet 70–2
air quality, and comfort 7
air-side economizers 183, 188
air speed, and comfort 36, 39
air temperature
  and comfort 35
  mixed-temperature sensor 72–3
  variations 39–40
all-air systems 24–5
  advantages 89
  by-pass box systems 94–5
  disadvantages 89–90
  dual-duct systems 95–8, 102

dual-duct variable air volume systems
  99–100, 102
dual path outside air systems 100–1
multizone systems 98–9, 102
reheat systems 24, 90–2, 101
three-deck multizone systems 99, 102
variable air volume (VAV) system 24–5,
  92–4, 102
all-water systems 25
analogue electronic controls 150
analogue input and output 158, 170
art work preservation, HVAC
  systems 5–6
ASHRAE (American Society of Heating,
  Refrigerating and Air-Conditioning
  Engineers)
  psychrometric chart 18–20
  Guideline 13-2000 *Specifying Direct
    Digital Control Systems* 169
  Standard 52.1-1992 *Gravimetric and Dust
    Spot Procedures for Testing Air
    Cleaning Devices* 49
  Standard 52.2-1999 *Method for Testing
    General Ventilation Air Cleaning
    Devices for the Removal Efficiency by
    Particle Size* 49–52, 111
  Standard 55-2004 *Thermal Environmental
    Conditions for Human Occupancy*
    32–3, 35–42, 109
  Standard 62 *Ventilation for Acceptable
    Indoor Air Quality* 52, 59
  Standard 62.1-2004 *Ventilation for
    Acceptable Indoor Air Quality* 44, 47,
    52–6, 59, 209
  Standard 62.2-2004 *Ventilation and
    Acceptable Indoor Air Quality in Low
    Rise Residential Buildings* 44, 52

ASHRAE *(Continued)*
  ASHRAE/IESNA Standard 90.1-2004
    *Energy Standard for Buildings Except Low-Rise Residential Buildings* 3, 161–2, 171, 176–9, 187–8
  Standard 135-2004 *A Data Communication Protocol for Building Automation and Control Networks* 166–7
axial fan 75

BACnet 166–7
bag filter 50, 52
barometric dampers 187
boilers
  central plant 136–9, 146
  condensing boiler 179
  replacement 175–6
  steam boilers 139
  two boiler system 138–9, 146
boreholes 204–5
breathing zone 55
building design
  and air-conditioning 26–7
  and energy conservation 172–6, 177
bypass box systems 94–5, 102
bypass damper 93–4

carbon dioxide 56–8
carbon monoxide 45
carcinogens 46
ceiling plenum 94
ceilings, radiant heating and cooling 109, 194
central plants 133–47
  boilers 133–4, 136–9, 146
  chillers 133–4, 139–42, 146
  comparison with local plants 134–5, 145–6
  cooling towers 142–5, 147
centrifugal compressor 140
centrifugal fan 75
changeover system, fan coils 111
chilled water, storage 197–201, 214
chilled water system 127–9, 132, 141–2, 175
chillers
  central plant 127–9, 139–42, 146
  energy efficiency 176, 178

client issues 28
climate
  and dual-path systems 100–1
  and economizer cycle 21–3, 31
  effects of 4
  and single zone air-handlers 72–3, 75
  and thermal storage 195
  and zones 62–5
closed loop controls 152–4, 169
closed water circuit 130–1, 132
clothing, and comfort 8, 33–4
Coanda effect 207
Coefficient of Performance (COP) 177, 178
comfort
  and environment 6–8, 9
  and indoor air quality 45, 47, 59
  *see also* thermal comfort
comfort cooling *see* air-conditioning systems
comfort envelope 36
compressor, in refrigeration equipment 75, 140
computers *see* Direct Digital Controls (DDC)
condensate, steam systems 118–19
condenser, in refrigeration equipment 75
condenser water 129–30, 134, 139–42
condensing boiler 179
contaminants
  filtration 48–52
  health effects 45–7
  indoor air quality 44–5, 58–9
  source control 47–8
controlled device 153
controlled variable 152
controller 153
control logic 73, 157
controls
  basics 150–5, 169
  choice of 149
  closed loop 152–4, 169
  Direct Digital Controls (DDC) 150, 157–69
  economizers 183
  electric 149
  electronic 150
  languages 166
  open loop 154
  pneumatic 149–50
  self-powered 149
  thermal storage 204
  time control 155–6

# Index

convection heating 104–8, 115
cooling
    evaporative cooling 185–6, 188–9
    radiant cooling 108–9, 194, 213
cooling coil 17–18, 21, 74
cooling towers 77–8, 129–30, 132, 134, 142–5, 147
costs 172

dampers
    in air-conditioning system 20, 71–2
    bypass damper 93–4
data gathering panel (DGP) 168
dead band 108
decoupled outdoor air systems 211–13, 215
dehumidification 101
desiccant wheels 182
dew point temperature 12
digital/binary input and output 158, 170
dilution ventilation 52
Direct Digital Controls (DDC) 150, 157–69, 204
    inputs and outputs 158–9
    naming conventions 159, 164
    sequence of operations 159, 162–4, 170
    single-zone air handlers 161–5
    system architecture 167–9
direct evaporative cooling 185, 188
disease, and air quality 46
displacement ventilation 208
drafts 39
drift eliminators 143
dual-duct systems 95–8, 102
dual-duct variable air volume systems 99–100, 102
dual-path outside air systems 100–1, 102, 211–13
dust spot efficiency 49

economizers
    air-side economizers 183, 188
    economizer cycle 21–3, 31, 89
    water-side 183–4, 188
electric controls 149
electricity, costs 196–7
electric thermal storage (ETS) heater 195–7
electronic controls 150

electronic filter 50
energy conservation
    air-side economizers 183, 188
    ASHRAE/IESNA Standard 90.1 3, 176–9, 187–8
    building design 171–6
    building pressure control 186–7, 189
    evaporative cooling 185–6, 188–9
    heat recovery 179–82
    water-side economizers 183–4, 188
energy-cost budget method 178–9, 188
energy efficiency, hot water systems 125–7
Energy Efficiency Ratio (EER) 177–8
enthalpy 14–15, 30, 76–7
entrained air 112
environment, for human comfort 6–8, 9
evaporative cooling 185–6, 188–9
evaporator 76
expansion valve 76
expectations, and comfort 8, 34–5

fan, in air-conditioning system 21, 75
fan coils 109–11, 115–16
farm animals, HVAC systems 5
filters 20, 48–52, 73
firing rate 138
float and thermostatic steam trap 119
floors
    radiant floor 108–9, 125, 191, 192
    surface temperature 40
four-pipe system, fan coil 111
frozen food storage, HVAC systems 5

gain 151–2, 169
greenhouse gas emissions 3
ground, heat source and sink 113, 204–6, 214

head
    pressure 184
    water flow 123
health, and air contaminants 45–7, 59
heating
    hydronic systems 103–16
    psychrometric chart 15
    radiant heating 191–3, 213
    steam systems 104, 118–20, 131
    water systems 120–4, 131

heating coil
   in air-conditioning system 21, 73–4
   fan coils 109–11, 115–16
heat pipes 180–2
heat pumps
   air-to-air system 78–80
   closed loop systems 114–15
   ground source heat pumps 113
   water source heat pumps 113–15, 116
heat recovery 136, 179–82, 188
   desiccant wheels 182, 188
   heat pipes 180–2, 188
   run-around energy recovery coils 179–80, 188
HEPA filter 52
hospitals
   ceiling panel heating 109
   dual-duct systems 96
   filters 51
   HVAC systems 5
hotels
   expectations 8, 35
   four-pipe fan-coil system 111
   ventilation 54–5
hot-water fan coils 111
hot water systems 124–7, 132
   boilers 136–9, 146
   energy efficiency 125–7, 175
human comfort *see* comfort
humidification, psychrometric chart 15–17
humidifier, in air-conditioning system 21, 74
humidistat 21, 67
humidity
   and comfort 35–6
   dehumidification 101
   relative humidity 12–15
   and zones 65, 67
humidity ratio (W) 11
HVAC (Heating, Ventilating and Air Conditioning)
   history of 2–3, 8–9
   system objectives 4–6, 9
hydronic circuits 117–18
hydronic systems 103–16
   advantages 104
   architecture of 117–32
   control of 106–7
   disadvantages 104
   fan coils 109–11, 115–16
   natural convection and low temperature radiation systems 104–8, 115
   panel heating and cooling 108–9, 115
   steam piping systems 118–20, 131
   two pipe induction systems 112, 116
   and ventilation 107–8
   water piping systems 120–30, 131
   water source heat pumps 113–15, 116

ice, storage 202–4, 214
IESNA (Illuminating Engineering Society of North America) *see* ASHRAE/IESNA *Standard 90.1-2004*
indirect evaporative cooling 185–6, 188–9
individuals, and comfort 8, 40
indoor air quality (IAQ) 3, 43–4, 58–9
   contaminants 44–7
   dilution 52
   filtration 48–52
   source control 47–8
   ventilation 52–8
induction, two pipe induction systems 112, 116
Induction Reheat Unit 90–2
infiltration 14
Integrated Part-Load Value (IPLV) 177, 178
Internet 168
interoperability 167

languages, controls 166
latent heat 14, 15, 30, 81
latent heat of fusion 198–9
Leadership in Energy and Environmental Design (LEED) 176, 188
legionella 46, 130, 145, 195
lighting
   and comfort 7
   energy conservation 174–5, 178
   and HVAC 3
low-grade heat 179
Low-Temperature Reheat Unit with Induced Air 90, 91, 92

mechanically conditioned spaces, comfort conditions 37–9
MERV (Minimum Efficiency Reporting Values) 49–52, 111

mixed temperature sensor, in air-conditioning system 72–3
mixing chamber, in air-conditioning system 20
modulating controls 151–4
mold, control of 47–8
multiple zone air systems 88–102
multizone systems 98–9, 102

offset 151–2, 169
on-off controls 150–1
on-off input and output 158
open loop control system 154, 169
open water circuit 130, 132
outdoor air, dual-path system 100–1, 211–13, 215
outdoor reset 107, 125, 154–5
outside air damper, in air-conditioning system 20, 71–2
overshoot 151–2, 169

panel filter 49–50
panel heating and cooling 108–9, 115, 191, 194
passive thermal storage 194–5, 214
personal environment model 6
physical space, and comfort 7
piping, water systems 120–3
pleated filter 50, 52
pneumatic controls 149–50
pollutants *see* contaminants
ponding, steam systems 120
pressure
　building pressure 186–7, 189
　and zones 65
proportional control 151–2, 169
protocols 166
psychosocial situation 7
psychrometric chart 11–20, 30
　acceptable temperature and humidity 38
　cooling coil 18
　cooling towers 144, 147
　design of 11
　evaporative cooling 185
　heating 15
　humidification 15–17
　relative humidity 12–15
pump curve 123, 126

pumps
　hot water systems 126–7
　water systems 123–4

radiant cooling 108–9, 194
radiant floor 108–9, 125, 191, 192
radiant heating
　high temperature 191–3, 213
　low temperature 104–8, 115, 191
radiant temperature 35, 40
radiators, heating system 105–6, 115
radon 46
reciprocating compressor 140
recuperator 136
refrigerant-based systems 26
refrigeration
　equipment 75–80, 86–7
　history of 2
　*see also* chillers
reheat system 24, 90–2, 101
　Induction Reheat Unit 90–2
　Low-Temperature Reheat Unit with Induced Air 90, 91, 92
relative humidity 12–15
relief air 72
reset
　chilled water 175
　heating 175
　outdoor reset 107, 125, 154–5
reset controller 155
return fan 75
rooftop units 82–5, 87
room air distribution systems 207–11, 214
run-around energy recovery coils 179–80, 188

safety issues, steam systems 119–20, 136–7
saturation line 12
saturation point 12
seasonal efficiency 135
secondary air *see* entrained air
self-powered controls 149
sensible heat 14, 81
sensor 153
setpoint 152
setpoint temperature 62, 154–5
sick building syndrome 47

single-zone air handlers 68–87
  components 70–5, 86
  direct digital control (DDC) 161–5
solar gain 62–4
solar heating, water 195
spaces
  attributes for comfort 7
  and zones 60–1
speed of reaction 169
split systems 85, 87
standalone panel 159
static lift 130
steam systems 104, 118–20, 131
  boilers 139, 146
  safety issues 119–20, 136–7
steam traps 118–19
storage heater 195–7
stratified tank 199–200
system choice matrix 28–30, 31
system curve 123
system head 123

Task/Ambient Conditioning system (TAC) 210
temperature *see* air temperature; radiant temperature
thermal comfort
  conditions for 7, 36–9, 41
  definition 32–3, 41
  factors 33–6, 41
  non-ideal conditions 39–40, 41
  non-standard groups 40, 42
thermal storage 190–1, 194–204, 214
  active 195–7
  chilled water storage 197–201
  controls 204
  ice storage 202–4
  passive 194–5
thermal variation, zones 64
thermostatic steam trap 118
thermostats 61, 65–7, 108, 192–3, 206
three-deck multizone systems 99, 102
time control 155–6
timing, and zones 64–5
tobacco smoke 46
transducer 158
turn-down ratio 138
Turn it off, Turn it down, Turn it in 174–6, 187
two pipe induction systems 112, 116

Under Floor Air Distribution (UFAD) 208–10
unitary refrigerant-based systems 26

variable air volume (VAV) systems 24–5, 92–4, 102
  controls 156–7
  direct digital control (DDC) 160–1
  dual-duct system 99–100, 102
variable input and output 158
ventilation
  acceptable indoor air quality 52–8
  air distribution 207–11, 214
  and hydronic heating systems 107–8
  occupant-operated windows 36–7, 107, 191, 206–7, 214
  zones 64
vertical temperature difference 39

water heating, passive 195
water piping systems 120–4, 131
  chilled water systems 127–9, 132, 146
  condenser water 129–30
  hot water systems 124–7, 132
water-side economizers 183–4, 188
water source heat pumps 113–15, 116
water systems *see* hydronic systems
water vapor, humidity ratio 11
web server 168
wells 204
window air-conditioners 4, 26, 76–8, 107, 177–8
windows
  and energy conservation 172–3, 174, 177
  occupant-controlled 36–7, 107, 191, 206–7, 214
  and zones 61–2

zone air distribution effectiveness 55
zoned air-conditioning systems 23–6, 31
  all-air systems 24–5, 89–102
  *see also* single-zone air handlers
zones
  control of 65–7
  definition 61, 67
  design 62–5